ADVANCES IN PROTEIN CHEMISTRY AND STRUCTURAL BIOLOGY

Volume 78

ADVANCES IN PROTEIN CHEMISTRY AND STRUCTURAL BIOLOGY

EDITED BY

ALEXANDER McPHERSON
University of California, Irvine
Department of Molecular Biology and Biochemistry
USA

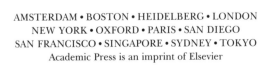

AMSTERDAM • BOSTON • HEIDELBERG • LONDON
NEW YORK • OXFORD • PARIS • SAN DIEGO
SAN FRANCISCO • SINGAPORE • SYDNEY • TOKYO
Academic Press is an imprint of Elsevier

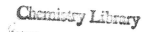

Academic Press is an imprint of Elsevier
The Boulevard, Langford Lane, Kidlington, Oxford OX5 1GB, UK
30 Corporate Drive, Suite 400, Burlington, MA 01803, USA
525 B Street, Suite 1900, San Diego, CA 92101-4495, USA

First edition 2009

ISBN: 978-0-12-374827-0
ISSN: 1876-1623

For information on all Academic Press publications
visit our website at www.elsevierdirect.com

Printed and bound in USA
09 10 11 12   10 9 8 7 6 5 4 3 2 1

Working together to grow
libraries in developing countries

www.elsevier.com | www.bookaid.org | www.sabre.org

ELSEVIER   BOOK AID
International   Sabre Foundation

Contents

Engineered Tropoelastin and Elastin-Based Biomaterials

STEVEN G. WISE, SUZANNE M. MITHIEUX, AND ANTHONY S. WEISS

The Architecture of the Cornea and Structural
Basis of Its Transparency

CARLO KNUPP, CHRISTIAN PINALI, PHILIP N. LEWIS, GERAINT J. PARFITT,
ROBERT D. YOUNG, KEITH M. MEEK, AND ANDREW J. QUANTOCK

Structural Biology of Periplasmic Chaperones

WILLIAM J. ALLEN, GILLES PHAN, AND GABRIEL WAKSMAN

Separate Roles of Structured and Unstructured Regions of Y-Family DNA Polymerases

Haruo Ohmori, Tomo Hanafusa, Eiji Ohashi, and Cyrus Vaziri

ENGINEERED TROPOELASTIN AND ELASTIN-BASED BIOMATERIALS

By STEVEN G. WISE, SUZANNE M. MITHIEUX, AND ANTHONY S. WEISS

School of Molecular and Microbial Biosciences G08, University of Sydney, Sydney, NSW 2006, Australia

Abstract

Elastin is a key mammalian extracellular matrix protein that is critical to the elasticity, compliance, and resilience of a range of tissues including the vasculature, skin, and lung. In addition to providing mechanical integrity to tissues, elastin also has critical functions in the regulation of cell behavior and may help to modulate the coagulation cascade. The high insolubility of elastin has limited its use to researchers, while soluble derivatives of elastin including elastin peptides, digested elastins, and tropoelastin have much broader applications. Recombinantly produced tropoelastin, the soluble monomer of elastin, has been shown to exhibit many of the properties intrinsic to the mature biopolymer. As such, recombinant human tropoelastin provides a versatile building block for the manufacture of biomaterials with potential for diverse applications in elastic tissues. One of the major benefits of soluble elastins is that they can be engineered into a range of physical forms. As a dominant example, soluble elastins including tropoelastin can form hydrogels when they are

1

chemically cross-linked. These self-organized constructs swell when transferred from a saline to aqueous environment and are highly elastic; these tunable responses are dependent on the types of cross-linker and elastin used. Soluble elastins can also be drawn into fine fibers using electrospinning. The morphology of these fibers can be altered by modifying spinning parameters that include delivery flow rate and the starting protein concentration. The resulting fibers then accumulate to form porous scaffolds, and can be wound around mandrils to create conduits for vascular applications. Electrospun scaffolds retain the elasticity and cell-interactive properties inherent in the tropoelastin precursor. Additionally, soluble elastins serve as versatile biomaterial coatings, enhancing cellular interactions and modulating the blood compatibility of polymer- and metal-based prostheses. Soluble elastins, and in particular tropoelastin, have highly favorable intrinsic physical and cell-interactive properties, warranting their adaption through incorporation into biomaterials and modification of implantable devices. The multiple choices of ways to produce elastin-based biomaterials mean that they are well suited to the tailoring of elastic biomaterials and hybrid constructs.

I. Introduction

Elastin is responsible for the elasticity of multiple tissues in all vertebrates apart from the cyclostomes, imparting resilience and recoil to elastic fibers. It is one of the most hydrophobic proteins currently known, with over 75% of the sequence made up of just four nonpolar amino acids; glycine, valine, alanine, and proline. Elastin is prevalent in several tissues, constituting up to 57% of the aorta, 50% of elastic ligaments, 32% of major vascular vessels, 7% of lung, and 5% of the dry weight of skin (Vrhovski and Weiss, 1998). Elastin is an extremely durable and insoluble biopolymer that does not turn over appreciably in healthy tissue, with a half-life of ~70 years (Petersen *et al.*, 2002).

Elastin is formed through the lysine-mediated cross-linking of its soluble precursor tropoelastin, which is a 60–72 kDa alternatively spliced extracellular matrix (ECM) protein. *In vivo*, tropoelastin is secreted from diverse elastogenic cell types including smooth muscle cells, fibroblasts, and endothelial cells. The deposited elastin is generically referred to in the elastin literature as the elastic fiber, although it should be appreciated that this is a broad term that is intended to encompass elastin in other forms, including elastin in the lamellar vascular wall. In mammals, it is deposited in the late fetal and early neonatal periods, after which tropoelastin production drops precipitously as there is very little turnover of mature elastin, so at maturity the production of new elastin ceases (Swee *et al.*, 1995). In the event of injury to elastic fibers, the repair production of tropoelastin can be quickly restarted and is influenced by a exogenous factors that include tumor necrosis factor-α (Kahari *et al.*, 1992), interleukin 1β (Kuang and Goldstein, 2003), insulin-like growth factor-1 (Jensen *et al.*, 1995), and particularly by transforming growth factor (Parks, 1997).

An important property of tropoelastin, in terms of elastic fiber formation, is its ability to self-assemble through coacervation. Coacervation has been well studied and is characterized by an inverse temperature transition and phase separation then fusion of approximately micron-sized particles of assembled protein molecules (Cox *et al.*, 1973a; Starcher *et al.*, 1973). Hydrophobic interactions between molecules at 37 °C, physiological salt, and pH 7–8 force tropoelastin out of solution as a viscous aggregate. The process can be monitored by observing the change in turbidity as a function of temperature (Urry and Long, 1977). The key outcome of coacervation is the alignment and concentration of tropoelastin prior to lysyl oxidase-assisted cross-linking (Kagan and Li, 2003).

Research into elastin and tropoelastin was initially restricted by the difficulty in obtaining these materials from animal sources, accentuated by the extreme insolubility of massively cross-linked elastin (Lansing *et al.*, 1952) and the short *in vivo* lifetime of tropoelastin which is quickly incorporated into elastin (Mecham and Foster, 1979). Overcoming these concerns, recombinant tropoelastin (rTE) was first expressed as a fusion protein in an *Escherichia coli* bacterial system after an expression vector containing the human cDNA sequence was constructed (Indik *et al.*, 1990). The high purity of the rTE produced in this system was beneficial; however, persistent degraded fragments and small yields were problematic. Following the optimization of a bacterial expression system using a synthetic elastin (SE) gene construct, a recombinant 60 kDa mature form of tropoelastin was expressed (Martin *et al.*, 1995). Significantly higher yields were achieved following several molecular biology modifications including the substitution of expression-limiting rare codons in the tropoelastin sequence with those more regularly found in the bacteria. This recombinant human tropoelastin (rhTE) is identical to the naturally secreted common form of human tropoelastin, is recognized and can be used by mammalian cells to form elastin (Stone *et al.*, 2001), associates at 37 °C (Toonkool *et al.*, 2001) and can be cross-linked *in vitro* both enzymatically (Mithieux *et al.*, 2005) and chemically (Wise *et al.*, 2005) to form an elastin-like material (Mithieux *et al.*, 2004). Refinement of purification and expression protocols now sees the monomer reproducibly produced in multigram quantities. The implicit advantages of rhTE are its relative availability, high purity and the philosophy of using the full-length molecule that provides for the requisite cell–molecular, intermolecular, and intramolecular interactions.

For biomaterials applications, tropoelastin is a preferred elastic construction component, as it mimics elastin's physical and biological properties and has the potential to replace damaged elastin-rich tissue. Engineered soluble elastins, including rhTE, α-elastin, and recombinantly produced elastin sequence-based variants such as elastin-like polypeptides, elastin-like proteins, and recombinant human elastin polypeptides, have the potential to augment tissues with a requisite elastin content such as skin dermis and vasculature. Potential dermal substitutes seek to take advantage of the natural elasticity, promoted cellular interactions, and opportunities for enhanced tissue regeneration as observed for soluble elastin-based biomaterials. As a vascular substitute, these biomaterials benefit from being able to support new matrix synthesis and to be endothelialized, while imparting some physical strength and particularly bestowing the recoil required of these vessels (Wise and Weiss, 2009).

rhTE provides the greatest versatility for biomaterials applications as it can be engineered to be suitable for a range of applications. Depending on the method of manufacture, rhTE can form elastic hydrogels and fine microfibers, and can be used as a surface coating. In these and other examined forms, rhTE constructs retain the key properties of elastin such as elasticity, enhanced cellular interactions, and blood compatibility which make it conveniently suited to biomaterial development. We describe here some key methods for the bioengineering of soluble elastins with a specific focus on rTE, outlining areas where it is likely to generate the greatest impact in the next generation of biomaterials.

II. HYDROGELS

Highly porous, interconnected, biocompatible scaffolds are particularly desirable for tissue engineering as they allow enhanced nutrient and oxygen transfer, cell migration, and proliferation. Polymer hydrogels are highly attractive for this purpose due to their hydrophilicity and high permeability (Annabi et al., 2009a). Hydrogels are cross-linked, water-swollen polymer networks that have great technological importance as biomedical materials (McMillan and Conticello, 2000). Researchers have long recognized the benefits of incorporating elastin and its derivatives into hydrogels, employing elastin polypeptides, acid-solubilized α-elastin, and ultimately rhTE. These hydrogels can be produced with a range of porosities and mechanical properties, while supporting the growth of a number of diverse cells types.

A. Elastin Sequence-Based Hydrogels

Dominating over 75% of its content with just four nonpolar amino acids (Gly, Val, Ala, Pro), the amino acid sequence of tropoelastin contains many repeating motifs. Typical examples include sequences such as VGVPG (Sandberg et al., 1986) and VPGVG or VGGVG (Li and Daggett, 2002). Synthetic polypeptides taking advantage of the intrinsic elasticity provided by these sequences have been widely studied and have strong potential for tissue engineering applications (Chilkoti et al., 2006). Individual domains of tropoelastin have also been produced, which have subsequently formed the basis of a minimalist approach to understanding the assembly of elastin. Specifically the relationship between structure and mechanical properties of elastin can be studied using reassembled tropoelastin domains, given that even very short sequences will self-assemble and can be cross-linked (Bellingham et al., 2001). Soluble elastin sequence-based molecules have been stabilized through the use of chemical cross-linkers including 1-ethyl-3-(3-dimethylaminopropyl) carbodiimide (EDC) and N-hydroxysuccinimide (NHS), glutaraldehyde (GA), bis(sulfosuccinimidyl) suberate (BS3), genipin (GP), pyrroloquinoline quinone (PQQ), and 1,6-diisocyanatohexane (HMDI). Cross-linking of these soluble elastins has led to the formation of a variety of hydrogels. (Mithieux et al., 2009).

Elastin-like polypeptides consisting of VPGXG repeats (where X was Lys every 7 or 17 pentapeptides, otherwise V), have been synthesized and chemically cross-linked to produce hydrogels. Gels ranged in stiffness from 0.24 to 3.7 kPa at 7 °C and from 1.6 to 15 kPa at 37 °C depending on protein concentration, lysine content, and molecular weight. Changes in gel properties suggest that at low temperatures, these structures are nearly completely elastic (Trabbic-Carlson *et al.*, 2003). Improvements to the mechanical and cell-interactive properties of these polypeptide materials has been achieved by copolymerizing sequence blocks from multiple parts of the elastin sequence, or combining with elements from other proteins. Constructs containing repeating elastin-derived (VPGIG)x repeating sequences, as well as cell-binding domains derived from fibronectin were cross-linked using glutaraldehyde. Tensile properties of cross-linked protein films were found to be inversely related to the molecular weights of the engineered constructs which varied from 14 to 59 kDa. At the highest cross-link density and lowest molecular weight, the elastic modulus was found to be similar to that of native elastin (Welsh and Tirrell, 2000).

The use of recombinant human elastin polypeptides has demonstrated that as few as three tropoelastin hydrophobic domains flanking two cross-linking domains are sufficient to support a self-assembly process that aligns lysines for cross-linking (Keeley *et al.*, 2002). These sequences contain sufficient information to self-organize into fibrillar structures and promote the formation of lysine-derived cross-links. When cross-linked with PQQ, these materials had an extensional elastic modulus of ~250 kPa, approaching the value commonly described for native elastin (300–600 kPa) (Fung, 1993). Sheets could be subjected to extensional strains of ~100% before breaking. The intrinsic ability of such polypeptides to self-organize into polymer structures not only makes them a useful model for understanding the process of assembly of elastin, but also sheds light on the design of self-assembling biomaterials (Bellingham *et al.*, 2003).

Recombinant human elastin polypeptides have also been cross-linked using GP, impacting on their physical and mechanical properties. The micron-scale topography of GP hydrogels revealed the presence of heterogeneity compared with PQQ analogs, which were comparatively uniform. It was also shown that the porosity of the GP hydrogels was much greater. GP-cross-linked sheets also exhibited significantly greater tensile strength, with a modulus greater than fourfold higher than similarly produced PQQ scaffolds. The change in physical characteristics appears to be caused by a higher cross-link density in the case of GP, likely due to its capacity to form both short- and long-range cross-links (Vieth *et al.*, 2007).

B. α-Elastin Hydrogels

As an alternative to tropoelastin or elastin-based peptides, some researchers artificially make "soluble elastin" by chemically treating animal derived elastin samples. The soluble elastins include α-elastin (Cox *et al.*, 1973b), an oxalic acid derivative of elastin, and κ-elastin, solubilized with potassium hydroxide. Using circular dichroism and Raman spectroscopy, the secondary structure of κ-elastin (Debelle *et al.*, 1995) was found to be similar to that for free monomer tropoelastin, supporting its use in a model system (Debelle *et al.*, 1998).

An elastic mimetic scaffold formed from α-elastin, cross-linked with a diepoxy cross-linker was produced in various pH conditions. Reaction pH was shown to modulate the degree of cross-linking, the swelling ratio, enzymatic degradation rate (8–50% per hour in 0.1 u/ml elastase), and elastic moduli (4–120 kPa) of these hydrogels. Cross-linking with a combination alkaline followed by neutral pH process resulted in materials with the highest degree of cross-links, as indicated by a low swelling, slow degradation rate, and high elastic modulus. Cross-linked α-elastin materials were subsequently shown to support vascular smooth muscle cell adhesion and a decreased proliferation rate compared to polystyrene controls (Leach *et al.*, 2005). They were also found to be fragile, as seen previously for other α-elastin-based materials (Bellingham and Keeley, 2004).

One means of improving the properties of α-elastin-based hydrogels has been to form the constructs in the presence of high-pressure carbon dioxide (CO_2). The dense gas process facilitates coacervation, improves cross-linking, and dramatically alters the micro- and macrostructures of pores within the sample. A complex network of interconnected pores was observed for GA cross-linked α-elastin hydrogels fabricated using high-pressure CO_2. *In vitro* cell culture studies demonstrated that these channels facilitated fibroblast penetration and proliferation within α-elastin structures (Annabi *et al.*, 2009b). However, the mechanical properties of these scaffolds was lower than anticipated, attributed to the low number of lysine residues (less than 1%) in α-elastin available for cross-linking with GA. Cross-linking of α-elastin with HMDI was performed in the presence of Dimethyl sulfoxide (DMSO) and gave rise to highly porous structurally stable α-elastin hydrogels. The increased solubility of CO_2 in DMSO compared to aqueous solution contributed to the fabrication of large pores and channels within the scaffolds as the dissolved CO_2 escaped on depressurization. This porosity substantially promoted cellular penetration and growth throughout the matrices. The highly porous α-elastin hydrogel structures fabricated in this study have potential for soft tissue engineering applications (Annabi *et al.*, 2009a).

C.   rhTE Hydrogels

rhTE has been cross-linked with both BS3 (Mithieux *et al.*, 2004) and GA (S. M. Mithieux and A. S. Weiss, unpublished data) to form SE hydrogel constructs. Using this approach, elastic sponges, sheets, and tubes have been formed (Fig. 1). These constructs have physical characteristics similar to those seen for native elastin. BS3-based constructs have

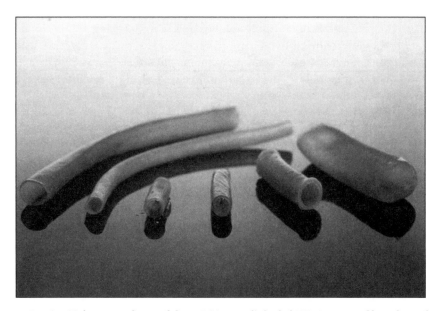

Fɪɢ. 1.   Tubes manufactured from BS3 cross-linked rhTE. A range of lengths and diameters are readily produced.

a Young's modulus ranging from 220 to 280 kPa and extensibility of 200–370%. GA cross-linked hydrogels had an average Young's modulus of 133 kPa and a mean elongation at failure of 234%. Like natural elastin, these SEs when dry are hard, brittle and inelastic but become elastic on wetting. They act as hydrogels and display stimuli-responsive characteristics toward temperature and salt concentrations, swelling under low salt and temperature conditions, and contracting at physiological salt and temperature. The manufacture of BS3 cross-linked hydrogels results in a smooth lower surface and a porous upper surface. Culture of human and murine adherent cells on the smooth surface leads to monolayer formation, while cells seeded on the open weave surface infiltrate into deeper compartments. Both BS3 and GA cross-linked SE hydrogels persist for at least 13 weeks following implantation into the dorsum of male guinea pigs where it was well tolerated, indicating that the material was innocuous and compatible *in vivo* (Mithieux *et al.*, 2004; S. M. Mithieux and A. S. Weiss, unpublished data).

The lysyl oxidase purified from the yeast strain *Pichia pastoris* (PPLO) can be used to enzymatically cross-link rhTE into hydrogel structures (Mithieux *et al.*, 2005). The process is considerably less efficient than when tropoelastin is oxidized *in vivo* by mammalian lysyl oxidases. Compared to elastin purified from horse ligamentum nuchae, PPLO-treated tropoelastin constructs contain ~10-fold more lysine residues and 25-fold less desmosine residues, indicating a decreased level of lysine oxidation. It is likely that PPLO is unable to sterically access all the lysine residues that would normally be oxidized by a mammalian LO (~30 kDa) due to the large size of this enzyme (~230 kDa). The decreased level of oxidation and subsequent cross-linking are reflected in the physical properties of the cross-linked tropoelastin. The Young's Modulus of the hydrogel is ~10 kPa, which is ~30–60-fold lower than that seen for elastin fibers (Fung, 1993). These values indicate a decreased stiffness in the PPLO-treated tropoelastin material, which is attributed to a decreased number of cross-links. In addition, the constructs swell following aqueous exposure to a much greater extent than normally seen for elastin further indicative of a loosely cross-linked matrix.

In a novel development, rhTE can also be assembled into an elastic biomaterial without the requirement for either enzyme- or chemical-mediated cross-linking. Under highly alkaline conditions, tropoelastin proceeds through a sol–gel transition leading to the formation of an irreversible hydrogel. The rhTE self-associates to give stable protein spheres that coalesce to generate solid integral hydrogels. This material supports human fibroblast proliferation *in vitro*, with cells colonizing the surface. Injection of the pH-modified rhTE intradermally into female Sprague–Dawley rats results in rapid localized gelation to form a persistent mass. The resulting elastic deposit persists for at least 2 weeks, encouragingly eliciting just a mild foreign body response and promoting collagen deposition and encapsulation of the material (Mithieux *et al.*, 2009).

Hydrogels made from soluble elastins can be produced with a wide range of physical properties, depending on the elastin starting material and choice of cross-linker. Material stiffness, strength, and porosity can be tailored by using varying sequence lengths and types, creating hydrogels in high-pressure environments and modulating reaction pH. Independent of these parameters, soluble elastin hydrogels importantly maintain the ability to support a range of cell types, and exhibit persistent elasticity. Biomaterials applications for elastin-based hydrogels include replacement of soft tissues, skin repair, and tissue bulking.

III. ELECTROSPUN MATERIALS

Electrospinning has recently emerged as a favored technique for generating biomimetic scaffolds made of synthetic and natural polymers for tissue engineering applications. It allows for the fabrication of multi-layered polymer-based scaffolds inspired by the natural architecture of the ECM (Han and Gouma, 2006). As a result, electrospinning of key ECM proteins such as collagen and elastin has become particularly prevalent.

Electrospinning of ECM proteins can generate fibers with diameters in the range from several micrometers down to less than 100 nm, which have a very high surface area to mass ratio. Accumulated fibers can be collected *n*-modifiable orientations to create three-dimensional (3-D) scaffolds with adjustable porosity that potentially present proteins in a form that allows their natural biological cues to be recognized by cells. In addition, electrospun fibers can retain the mechanical properties inherent in the source protein as has been demonstrated for both collagen and elastin (Boland *et al.*, 2004). In this way, biomaterial scaffolds are fabricated by electrospinning to exhibit favorable mechanical properties, facilitating cell attachment, cell growth, and regulating cell differentiation.

In a typical electrospinning setup, high-voltage fields cause polymers in volatile solvents to elongate and splay into small fibers. The fibers are drawn to a grounded surface and/or adhere to surfaces placed between the solution source and ground (Stitzel *et al.*, 2006). The morphology of the fibers collected at the plate as a nonwoven mesh is influenced by several parameters, particularly the solvent used, the potential difference of the applied electric field, the flow rate, and the collecting distance (Buttafoco *et al.*, 2006).

A. Elastin Sequence-Based Electrospinning

Elastin sequence-based fibers were first produced using an 81-kDa elastin-like protein based upon the repeat sequence VGPVG (Huang *et al.*, 2000). Electrospinning of solutions above 10% (w/v) resulted in long uniform fibers. An average fiber diameter of 450 nm was observed for this protein construct with generally a flat ribbon-like morphology. Optimal fiber formation was observed with use of an 18-kV electric field and a 15-cm distance between the spinneret and plate collector. Concentration of the polymer solution and the delivery flow rate were found to have the most effect on final fiber morphology. These early findings established many of the basic principles for the successful electrospinning of elastin-based polymers, investigated further by subsequent researchers.

High-molecular-weight elastin-like polypeptides have also been electrospun. These constructs were based on hydrophobic IPAVG end blocks, separated by VPGVG-repeat mid-block elements. When dissolved in 2,2,2-trifluoroethanol (TFE), electrospinning proved to be a feasible strategy for creating protein fibers with diameters ranging from 100 to 400 nm (Nagapudi *et al.*, 2005). Spinning of the same polypeptides from water results in fibers with diameters ranging from 800 to 3 mm. At identical concentrations polypeptide solutions in TFE displays a lower viscosity at 23 °C than aqueous protein solutions at 5 °C, which may contribute to the formation of smaller diameter fibers.

B. α-Elastin and Composites

Soluble α-elastin is useful for electrospinning biomaterial scaffolds. Initial investigations of spinning parameters such as protein concentration, flow rate, and applied voltage were carried out by Li *et al.* (2005). The width of electrospun α-elastin fibers depends primarily on the concentration of the solution. Electrospinning of α-elastin at 10% (w/v) yield large beads and fragmented fibers. At 15%, there are no beads, but the fibers are fragmented, while at 20% the electrospun fibers are continuous and uniform. Fibers electrospun at 20% concentration are also wider than those at 10% and 15%. On increasing the delivery rate from 1 to 8 ml/h, the mean width of the fibers increases from 0.6 ± 0.1 to 3.6 ± 0.7 μm. Remarkably, following fractionation of the oxalic acid digest, based on cross-link content, the α-elastin fractions can still be electrospun to give fibers of similar dimensions, ranging from 1 to 10 μm but fibers spun from the fraction containing the highest proportion of cross-links has an enhanced signaling effect on vascular smooth muscle cells, possibly due to the increased presentation of receptor sequences like VGVAPG in the cross-linked 3-D elastin. After cross-linking and hydration, the fibers display elasticity, where they can elongate 200% (Miyamoto *et al.*, 2009).

The versatility of the electrospinning technique makes it possible to combine multiple polymers and produce scaffolds with mechanical and biological properties superior to those of the individual components. This blending approach is particularly useful for mimicking complex, multicomponent tissues such as the vasculature. The use of electrospun α-elastin in candidate vascular compatible scaffolds has historically focused on the use of composites. In this respect, an electrospun vascular graft composed of polydioxanone (PDO) and α-elastin has been produced with the aim of matching the mechanical properties of native arterial tissue, where PDO conveys mechanical integrity, while elastin provides elasticity and bioactivity. The elasticity of α-elastin dominates the low-strain mechanical response of the vessel to blood flow. At equal ratios of elastin and PDO, the resulting grafts most closely mimic the compliance of native femoral artery, but overall strength is lower than anticipated. Additional reinforcement of the grafts with wound Prolene sutures partly alleviates this concern (Smith *et al.*, 2008). Preliminary cell culture studies show the α-elastin-containing grafts to be bioactive as cells migrate through their full thickness within 7 days, but fail to migrate into pure PDO scaffolds (Sell *et al.*, 2006).

Similarly, researchers have combined the two most abundant proteins in a native blood vessel, collagen, and elastin, in electrospun constructs. Spinning collagen/α-elastin solutions yields meshes composed of fibers 220–600 nm in diameter, with spinning enhanced by the addition of polyethylene oxide (PEO) and sodium chloride (NaCl) to the solution. After EDC/NHS cross-linking, the PEO and NaCl are washed out of the final scaffolds. Although detailed mechanical data are not available, vascular smooth muscle cells grow on the collagen/elastin hybrid scaffolds to give a confluent layer after 14 days, supporting the potential of this category of materials for vascular repair (Buttafoco *et al.*, 2006).

With an increase in the complexity of synthetic grafts scaffolds, elastin has also been coblended with collagen and a third biodegradable synthetic polymer, such as in the formation of vascular graft scaffolds that utilize electrospun polymer blends of Type I collagen, α-elastin, and poly (D,L-lactide-co-glycolide) (PLGA). Addition of PLGA improves electrospinning in this regard in addition to mechanically strengthening constructs. Electrospun scaffolds can exhibit compliance similar to that of a native bovine iliac artery and a burst pressure of 1,425 mmHg. Additionally, the three-component hybrid grafts support EC and smooth muscle cell (SMC) growth, while subcutaneous implantation in mice indicates that the samples are only mildly inflammatory and nontoxic (Stitzel *et al.*, 2006). By changing the synthetic biodegradable polymer (PLGA, poly(L-lactide) (PLLA), poly-caprolactone (PCL), and poly(lactide-cocaprolactone)), mechanical properties can be further finely tuned. The tensile strength of a collagen/elastin/PLLA scaffold is elevated through blending (0.83 MPa), while a collagen/elastin/PCL scaffold is more elastic (Lee *et al.*, 2007).

C. rhTE Electrospinning

rhTE electrospinning was first carried out with the aim of fabricating tissue engineered scaffolds. In contrast to collagen and gelatin, which produce fibers at a nanometer scale, the diameter of tropoelastin fibers is reproducibly in the range of microns. Electrospun tropoelastin fibers are typically 2–3 times wider than those spun under the same conditions from α-elastin. These fibers are also flatter and appear wavy at high delivery rates, while collagen and gelatin fibers are typically straight. The innate elastic properties of tropoelastin are retained upon electrospinning. Tropoelastin constructs are more elastic than α-elastin, gelatin, or collagen fibers. Human embryonic palatal mesenchymal cells attach to, spread, migrate, and proliferate equally well on the elastic scaffolds, although tropoelastin is advantageous over α-elastin for fabricating these types of engineered scaffolds due to increased elasticity (Li *et al.*, 2005).

The physical characterization of electrospun proteins is often limited to cataloguing changes to fiber morphologies in response to modifying parameters such as flow rate and concentration. However, following reports that some ECM proteins such as collagen(s) lose their tertiary structure and degrade when electrospun (Zeugolis *et al.*, 2008), this aspect of rhTE electrospinning was of particular interest. Examination of the secondary structure, coacervation ability and SDS-PAGE profile of rhTE before and after electrospinning demonstrates that the protein is not adversely affected by this technique. Subsequent to these studies, rhTE microfibers (Fig. 2) that are cross-linked with either HMDI or GA were shown to support the attachment and growth of both vascular and skin cell types. Proliferation continues for at least 72 h *in vitro* where these cells are maintained on the scaffolds (Nivison-Smith *et al.*, 2009).

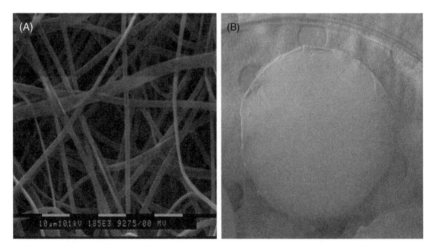

Fɪɢ. 2. Recombinant human elastin can be electrospun to form (A) fine microfibers that can in turn accumulate to form (B) scaffolds for biomaterials applications. (See Color Insert.)

In an advance in the use of electrospun rhTE toward the goal of clinical applications, a cell-interactive dermal substitute scaffold was recently described. This work is predicated on the knowledge that when coated on tissue culture plastic, rhTE promotes primary human dermal fibroblast attachment, spreading, and proliferation (Bax *et al.*, 2009b), which is a critical prerequisite in bioengineering scaffolds for skin repair. Employing electrospun rhTE, open-weave, fibrous scaffolds that closely mimic the fibrous nature of the skin dermis are robust and consist of flat, ribbon-like fibers with widths (average $2.3 \pm 0.5\,\mu$m) that are similar to those of native dermal elastic fibers (0.7–$5\,\mu$m). The scaffolds also display elasticity (265 ± 17 kPa) similar to that of natural elastin. These rhTE scaffolds interact with primary human dermal fibroblasts, which consistently attach and proliferate to form monolayers across the scaffold surface. By altering these electrospinning conditions, an open weave design was formulated, thereby enlarging the spaces between individual fibers and giving greater fiber diameters that beneficially allow for a substantial increase in cell infiltration throughout the scaffolds (Rnjak *et al.*, 2009).

Clinical vascular applications are also within the gamut of electrospun rhTE. While rhTE alone has insufficient mechanical strength for grafting applications, cospinning with synthetic polymers such as PCL substantially enhances the construct's tensile strength. A vascular graft with elasticity matching that of the human internal mammary arteries has been produced on this basis. These grafts support the attachment and proliferation of endothelial cells and have low thrombogenicity *in vitro* with potential for use as small-diameter vascular grafts (Wise, S.G. *et al.*, in preparation).

Biomaterials scaffolds made from soluble elastins for a range of clinical applications can now be considered for manufacturing using electrospinning. Process parameters such as protein concentration and flow rate impact greatly on the morphology of the resulting fibers, allowing for the production of scaffolds of reproducibly adjusted elasticity, porosity, and strength. The addition of synthetic and other ECM protein copolymers to these scaffolds can further modify these characteristics in order to more closely mimic the native tissue that is to be replaced. rhTE is of particular interest for electrospinning given the high elasticity of the microfibers produced, coupled with the promising cell-interactive properties demonstrated in its use as a potential dermal substitute.

IV. MATERIAL COATING

A major measure of the success of devices that are implanted *in vivo* is the degree of early cellular responses. These responses are in part mediated by the concentration, composition, and conformation of adsorbed proteins at the implant surface (Nath *et al.*, 2004). For implants that are innately nonbiological such as stainless steel, one strategy to improve their interaction and biocompatibility is to coat them with biologically favorable agents. These agents can be used to modulate cell attachment and proliferation, fibrosis, and blood compatibility. Given the roles of elastins in cell signaling and its potential to limit blood coagulation, this is a promising candidate for the coating of implanted medical devices. A wide range of biological properties have been attributed to elastin, tropoelastin, and elastin-derived peptides. Elastin contains multiple signals for cell and protein attachment and can promote cellular chemotaxis. The modulation of various cell types has been ascribed primarily to two identified cell surface elastin receptors. A nonintegrin cell surface receptor complex containing a peripheral elastin binding subunit termed the elastin-binding protein EBP binds the VGVAPG elastin sequence with high affinity (Duca *et al.*, 2007; Hinek *et al.*, 1988; Robert *et al.*, 1992). A more recently identified cell adhesion receptor to tropoelastin mediates its effect through the C-terminal tropoelastin sequence GRKRK and cell surface integrin $\alpha_v \beta_3$ (Bax *et al.*, 2009b; Rodgers and Weiss, 2004). In addition, tropoelastin can interact with heparan and chondroitin sulfate-containing cell-surface glycosaminoglycans via its C-terminal domain (Broekelmann *et al.*, 2005). Cellular responses to intact tropoelastin encompass stimulation of vasodilation and intracellular Ca^{2+} mobilization. Elastin is also a potent autocrine regulator of SMC activity that inhibits proliferation and regulates migration (Li *et al.*, 1998). In a porcine coronary model of restenosis, delivery of exogenous elastin to injured vessels *in vivo* reduces neointimal formation (Karnik *et al.*, 2003). Biological activities attributed to elastin peptides include the regulation of proliferation, chemotaxis, protease upregulation, and promoted human fibroblast cell survival (Cantarelli *et al.*, 2009).

In addition to its well-known cell signaling properties, elastin and sequence derivatives interact favorably with blood and do not promote thrombosis. One of the primary drivers for this phenomenon is the relatively innocuous effect of elastin on platelets. Elastin elicits minimal platelet adhesion, degranulation, and aggregation (Baumgartner *et al.*, 1976). In contrast, other ECM proteins used in biomaterials such as collagens I and III and fibronectin induce aggregation of platelets and are accordingly considered thrombogenic (Barnes and MacIntyre, 1979; Seeger and Klingman, 1988). Platelets are of particular importance in this setting as it is recognized that platelet-mediated mechanisms rather than thrombin activation are the major contributors to the thrombosis of metallic prostheses (Schomig *et al.*, 1996). Applied clinically, the coating of surfaces with soluble elastins has translated to improved blood interactions and reduced clotting, potentially leading to better blood compatibility of cardiovascular devices such as vascular conduits and arterial/venous catheters.

A. Elastin Sequence-Based Coatings

Recombinant human elastin polypeptides have been used to coat commercial synthetic materials including polyethylene terephthalate, a poly(tetrafluoroethylene/ethylene) copolymer, and a polycarbonate polyurethane (Woodhouse *et al.*, 2004). Compared to noncoated controls, there is a significant decrease in both platelet microparticle release and P-selectin expression for the polypeptide-coated surfaces, indicating an overall reduction in platelet activation. Fewer platelets adhere to coated surfaces. In a rabbit model, evaluations of polyurethane catheters coated with the polypeptide reveal a marked increase in catheter patency and a significant decrease in fibrin accretion and embolism when compared to uncoated controls.

Similarly, a recombinant elastin-mimetic triblock protein polymer was used to impregnate small-diameter expanded polytetrafluoroethylene (ePTFE) vascular grafts, then blood contacting properties evaluated using a baboon extracorporeal femoral arteriovenous shunt model. This model is of particular interest given the close primate phylogenetic resemblance between the hemostatic systems of baboon and human. Minimal platelet deposition occurs on elastin-coated ePTFE graft surfaces with total adsorbed fibrinogen during the test period at $0.03 \pm 0.02 \, \mathrm{mg/cm^2}$ for the elastin-coated grafts compared to $1.44 \pm 0.75 \, \mathrm{mg/cm^2}$ on uncoated ePTFE grafts. These data point to the applicability of elastin-mimetic protein polymers as a viable nonthrombogenic coating (Jordan *et al.*, 2007).

B.   rhTE Coatings

Protein coatings can be applied to a range of substrates, providing that an appropriate surface modification is made to facilitate binding. For medically utilized polymers, plasma immersion ion implantation (PIII) treatment (Kondyurin and Bilek, 2008; Kondyurin *et al.*, 2008) activates otherwise inert substrates and predisposes them to covalently bind protein. Given that many prosthetic implants are made from polymers, it is desirable for these polymers to promote biological function by promoting the adhesion, differentiation and viability of cells. Binding of rhTE to polystyrene in this manner promotes fibroblast spreading only on the treated surfaces, indicating that rhTE renders the PIII-treated surface biologically active. When a contact mask is used to pattern the PIII treatment, cells preferentially spread on the tropoelastin coated, PIII-treated sections of the surface, demonstrating that PIII treatment of polystyrene improves the polymer's rhTE binding properties, with opportunities for tissue engineering and prosthetic design (Bax *et al.*, 2009a).

Applying this idea more broadly to nonpolymer surfaces would be very useful but metal has proved to be more difficult to surface-modify. Indeed, while binding proteins to polymers is relatively widespread, covalent binding of proteins to metallic surfaces has long been problematic and elusive. The recently developed plasma polymerization (PP) method (Yin *et al.*, 2009a) deposits an activated polymer surface suitable for protein binding. Using stainless steel as a substrate, PP allows for the attachment of rhTE. Tropoelastin covalently binds to the surface, supplemented with additional physisorbed multilayers on extended incubation. The physisorbed rhTE layers can be washed away in buffer or SDS, whereas a monolayer of covalently attached tropoelastin remained tightly bound. The plasma-coated stainless-steel surface with immobilized tropoelastin has improved biocompatibility, enhances endothelial attachment and proliferation by $86.3 \pm 10.5\%$ and $76.9 \pm 6.4\%$, respectively ($P < 0.01$ vs. control), relative to uncoated stainless-steel controls (Yin *et al.*, 2009b). Clinically, metal is used in many of the implants used in vascular repair due to their conveyed strength. However the performance of most vascular implants, in particular stents, is highly dependent on the degree and speed of re-endothelialization, given that areas of injury that rapidly re-endothelialize have significantly less intimal thickening and restenosis (Lau *et al.*, 2003), while also deterring thrombus formation (Wu and Thiagarajan, 1996). rhTE coatings applied to otherwise inert substrates using this technology could have broad applications to a range of nonpolymer vascular devices.

Soluble elastins influence both cell-interactive and blood-contacting properties when employed as biomaterial coatings. The nonthrombogenic nature of recombinant human elastin polypeptides and block polymers has been demonstrated *in vivo* and hold great promise in a range of clinical vascular applications. Covalently immobilized rhTE on both polymeric and stainless-steel surfaces enhances cell attachment and proliferation of skin and vascular cell types. Overall, the biocompatibility of both polymer and metal medical implants stands to be improved by coating with soluble elastin derivatives, and particularly with rhTE due to its ability to mimic many of the properties of natural elastin.

REFERENCES

Annabi, N., Mithieux, S. M., Boughton, E. A., Ruys, A. J., Weiss, A. S., and Dehghani, F. (2009a). Synthesis of highly porous crosslinked elastin hydrogels and their interaction with fibroblasts in vitro. *Biomaterials* **30**, 4550–4557.

Annabi, N., Mithieux, S. M., Weiss, A. S., and Dehghani, F. (2009b). The fabrication of elastin-based hydrogels using high pressure CO_2. *Biomaterials* **30**, 1–7.

Barnes, M. J., and MacIntyre, D. E. (1979). Platelet-reactivity of isolated constituents of the blood vessel wall. *Haemostasis* **8**, 158–170.

Baumgartner, H. R., Muggli, R., Tschopp, T. B., and Turitto, V. T. (1976). Platelet adhesion, release and aggregation in flowing blood: Effects of surface properties and platelet function. *Thromb. Haemost.* **35**, 124–138.

Bax, D. V., *et al.* (2009a). Linker-free covalent attachment of the extracellular matrix protein tropoelastin to a polymer surface for directed cell spreading. *Acta Biomater.*, **5**, 3371–3381, doi:10.1016/j.actbio.2009.05.016.

Bax, D. V., *et al.* (2009b). Cell adhesion to tropoelastin is mediated via the C-terminal GRKRK motif and integrin alphaVbeta3. *J. Biol. Chem.*, **284**, 28616–28623, doi/ 10.1074/jbc.M109.017525.

Bellingham, C. M., and Keeley, F. W. (2004). Self-ordered polymerization of elastin-based biomaterials. *Curr. Opin. Solid State Mater. Sci.* **8**, 135–139.

Bellingham, C. M., Lillie, M. A., Gosline, J. M., Wright, G. M., Starcher, B. C., Bailey, A. J., Woodhouse, K. A., and Keeley, F. W. (2003). Recombinant human elastin polypeptides self-assemble into biomaterials with elastin-like properties. *Biopolymers* **70**, 445–455.

Bellingham, C. M., Woodhouse, K. A., Robson, P., Rothstein, S. J., and Keeley, F. W. (2001). Self-aggregation characteristics of recombinantly expressed human elastin polypeptides. *Biochim. Biophys. Acta* **1550**, 6–19.

Boland, E. D., Matthews, J. A., Pawlowski, K. J., Simpson, D. G., Wnek, G. E., and Bowlin, G. L. (2004). Electrospinning collagen and elastin: Preliminary vascular tissue engineering. *Front. Biosci.* **9**, 1422–1432.

Broekelmann, T. J., Kozel, B. A., Ishibashi, H., Werneck, C. C., Keeley, F. W., Zhang, L., and Mecham, R. P. (2005). Tropoelastin interacts with cell-surface glycosaminoglycans via its C-terminal domain. *J. Biol. Chem.* **280**, 40939–40947.

Buttafoco, L., Kolkman, N. G., Engbers-Buijtenhuijs, P., Poot, A. A., Dijkstra, P. J., Vermes, I., and Feijen, J. (2006). Electrospinning of collagen and elastin for tissue engineering applications. *Biomaterials* **27**, 724–734.

Cantarelli, B., Duca, L., Blanchevoye, C., Poitevin, S., Martiny, L., and Debelle, L. (2009). Elastin peptides antagonize ceramide-induced apoptosis. *FEBS Lett.* **583**, 2385–2391.

Chilkoti, A., Christensen, T., and MacKay, J. A. (2006). Stimulus responsive elastin biopolymers: Applications in medicine and biotechnology. *Curr. Opin. Chem. Biol.* **10**, 652–657.

Cox, B. A., Starcher, B. C., and Urry, D. W. (1973a). Coacervation of tropoelastin results in fiber formation. *J. Biol. Chem.* **249**, 997–998.

Cox, B. A., Starcher, B. C., and Urry, D. W. (1973b). Coacervation of alpha-elastin results in fiber formation. *Biochim. Biophys. Acta* **317**, 209–213.

Debelle, L., Alix, A. J., Jacob, M. P., Huvenne, J. P., Berjot, M., Sombret, B., and Legrand, P. (1995). Bovine elastin and kappa-elastin secondary structure determination by optical spectroscopies. *J. Biol. Chem.* **270**, 26099–26103.

Debelle, L., Alix, A. J., Wei, S. M., Jacob, M. P., Huvenne, J. P., Berjot, M., and Legrand, P. (1998). The secondary structure and architecture of human elastin. *Eur. J. Biochem.* **258**, 533–539.

Duca, L., Blanchevoye, C., Cantarelli, B., Ghoneim, C., Dedieu, S., Delacoux, F., Hornebeck, W., Hinek, A., Martiny, L., and Debelle, L. (2007). The elastin receptor complex transduces signals through the catalytic activity of its Neu-1 subunit. *J. Biol. Chem.* **282**, 12484–12491.

Fung, Y. C. (1993). "Biomechanics: Mechanical Properties of Living Tissues." Springer-Verlag, New York.

Han, D., and Gouma, P. I. (2006). Electrospun bioscaffolds that mimic the topology of extracellular matrix. *Nanomedicine* **2**, 37–41.

Hinek, A., Wrenn, D. S., Mecham, R. P., and Barondes, S. H. (1988). The elastin receptor: A galactoside-binding protein. *Science* **239**, 1539–1541.

Huang, L., McMillan, R. A., Apkarian, R. P., Pourdeyhimi, B., Conticello, V. P., and Chaikof, E. L. (2000). Generation of synthetic elastin-mimetic small diameter fibers and fiber networks. *Macromolecules* **33**, 2989–2997.

Indik, Z., Abrams, W. R., Kucich, U., Gibson, C. W., Mecham, R. P., and Rosenbloom, J. (1990). Production of recombinant human tropoelastin: Characterization and demonstration of immunologic and chemotactic activity. *Arch. Biochem. Biophys.* **280**, 80–86.

Jensen, D. E., Rich, C. B., Terpstra, A. J., Farmer, S. R., and Foster, J. A. (1995). Transcriptional regulation of the elastin gene by insulin-like growth factor-I involves disruption of Sp1 binding. Evidence for the role of Rb in mediating Sp1 binding in aortic smooth muscle cells. *J. Biol. Chem.* **270**, 6555–6563.

Jordan, S. W., Haller, C. A., Sallach, R. E., Apkarian, R. P., Hanson, S. R., and Chaikof, E. L. (2007). The effect of a recombinant elastin-mimetic coating of an ePTFE prosthesis on acute thrombogenicity in a baboon arteriovenous shunt. *Biomaterials* **28**, 1191–1197.

Kagan, H. M., and Li, W. (2003). Lysyl oxidase: Properties, specificity, and biological roles inside and outside of the cell. *J. Cell. Biochem.* **88**, 660–672.

Kahari, V. M., Chen, Y. Q., Bashir, M. M., Rosenbloom, J., and Uittoet, J. (1992). Tumor necrosis factor-alpha down-regulates human elastin gene expression. Evidence for the role of AP-1 in the suppression of promoter activity. *J. Biol. Chem.* **267**, 26134–26141.

Karnik, S. K., Brooke, B. S., Bayes-Genis, A., Sorensen, L., Wythe, J. D., Schwartz, R. S., Keating, M. T., and Li, D. Y. (2003). A critical role for elastin signaling in vascular morphogenesis and disease. *Development* **130**, 411–423.

Keeley, F. W., Bellingham, C. M., and Woodhouse, K. A. (2002). Elastin as a self-organizing biomaterial: Use of recombinantly expressed human elastin polypeptides as a model for investigations of structure and self-assembly of elastin. *Philos. Trans. R. Soc. Lond. B Biol. Sci.* **357**, 185–189.

Kondyurin, A., and Bilek, M. M. (2008). "Ion Beam Treatment of Polymers." Elsevier Inc., San Diego.

Kondyurin, A., Nosworthy, N. J., and Bilek, M. M. (2008). Attachment of horseradish peroxidase to polytetrafluorethylene (teflon) after plasma immersion ion implantation. *Acta Biomater.* **4**, 1218–1225.

Kuang, P. P., and Goldstein, R. H. (2003). Regulation of elastin gene transcription by interleukin-1 beta-induced C/EBP beta isoforms. *Am. J. Phys. Cell Phys.* **285**, C1349–C1355.

Lansing, A. I., Rosenthal, T. B., Alex, M., and Dempsey, E. W. (1952). The structure and chemical characterization of elastic fibers as revealed by elastase and by electron microscopy. *Anat. Rec.* **114**, 555–575.

Lau, A. K., Leichtweis, S. B., Hume, P., Mashima, R., Hou, J. Y., Chaufour, X., Wilkinson, B., Hunt, N. H., Celermajer, D. S., and Stocker, R. (2003). Probucol promotes functional reendothelialization in balloon-injured rabbit aortas. *Circulation* **107**, 2031–2036.

Leach, J. B., Wolinsky, J. B., Stone, P. J., and Wong, J. Y. (2005). Crosslinked alpha-elastin biomaterials: Towards a processable elastin mimetic scaffold. *Acta Biomater.* **1**, 155–164.

Lee, S. J., Yoo, J. J., Lim, G. J., Atala, A., and Stitzel, J.. (2007). In vitro evaluation of electrospun nanofiber scaffolds for vascular graft application. *J. Biomed. Mat. Res. A.* **83**, 999–1008.

Li, B., and Daggett, V. (2002). Molecular basis for the extensibility of elastin. *J. Muscle Res. Cell Motil.* **23**, 561–573.

Li, D. Y., Brooke, B., Davis, E. C., Mecham, R. P., Sorensen, L. K., Boak, B. B., Eichwald, E., and Keating, M. T. (1998). Elastin is an essential determinant of arterial morphogenesis. *Nature* **393**, 276–280.

Li, M., Mondrinos, M. J., Gandhi, M. R., Ko, F. K., Weiss, A. S., and Lelkes, P. I. (2005). Electrospun protein fibers as matrices for tissue engineering. *Biomaterials* **26**, 5999–6008.

Martin, S. L., Vrhovski, B., and Weiss, A. S. (1995). Total synthesis and expression in Escherichia coli of a gene encoding human tropoelastin. *Gene* **154**, 159–166.

McMillan, R. A., and Conticello, V. P. (2000). Synthesis and characterization of elastin-mimetic protein gels derived from a well-defined polypeptide precursor. *Macromolecules* **33**, 4809–4821.

Mecham, R. P., and Foster, J. A. (1979). Characterization of insoluble elastin from copper-deficient pigs. Its usefulness in elastin sequence studies. *Biochim. Biophys. Acta* **577**, 147–158.

Mithieux, S. M., Rasko, J. E., and Weiss, A. S. (2004). Synthetic elastin hydrogels derived from massive elastic assemblies of self-organized human protein monomers. *Biomaterials* **25**, 4921–4927.

Mithieux, S. M., Tu, Y., Korkmaz, E., Braet, F., and Weiss, A. S. (2009). In situ polymerization of tropoelastin in the absence of chemical cross-linking. *Biomaterials* **30**, 431–435.

Mithieux, S. M., Wise, S. G., Raftery, M. J., Starcher, B., and Weiss, A. S. (2005). A model two-component system for studying the architecture of elastin assembly in vitro. *J. Struct. Biol.* **149**, 282–289.

Miyamoto, K., Atarashi, M., Kadozono, H., Shibata, M., Koyama, Y., Okai, M., Inakuma, A., Kitazono, E., Kaneko, H., Takebayashi, T., and Horiuchi, T.. (2009). Creation of cross-linked electrospun isotypic-elastin fibers controlled cell-differentiation with new cross-linker. *Int. J. Biol. Macromol.* **45**, 33–41.

Nagapudi, K., Brinkman, W. T., Thomas, B. S., Park, J. O., Srinivasarao, M., Wright, E., Conticello, V. P., and Chaikof, E. L. (2005). Viscoelastic and mechanical behavior of recombinant protein elastomers. *Biomaterials* **26**, 4695–4706.

Nath, N., Hyun, J., Ma, H., and Chilkoti, A. (2004). Surface engineering strategies for control of protein and cell interactions. *Surf. Sci.* **570**, 98–110.

Nivison-Smith, L., *et al.* (2009). Synthetic human elastin microfibers: Stable cross-linked tropoelastin and cell interactive constructs for tissue engineering applications. *Acta Biomater.*, in press, doi:10.1016/j.actbio.2009.08.011.

Parks, W. C. (1997). Posttranscriptional regulation of lung elastin production. *Am. J. Respir. Cell Mol. Biol.* **17**, 1–2.

Petersen, E., Wagberg, F., and Ängquist, K.-A. (2002). Serum concentrations of elastin-derived peptides in patients with specific manifestations of atherosclerotic disease. *Eur. J. Vasc. Endovasc. Surg.* **24**, 440–444.

Rnjak, J., *et al.* (2009). Primary human dermal fibroblast interactions with open weave three-dimensional scaffolds prepared from synthetic human elastin. *Biomaterials*, **30**, 6469–6477, doi:10.1016/j.biomaterials.2009.08.017.

Robert, L., Jacob, M. P., and Labat-Robert, J. (1992) Cell-matrix interactions in the genesis of arteriosclerosis and atheroma. Effect of aging. *Ann. N. Y. Acad. Sci.* **673**, 331–341.

Rodgers, U. R., and Weiss, A. (2004). Integrin avb3 binds a unique non-RGD site near the C-terminus of human tropoelastin. *Biochimie* **86**, 173–178.

Sandberg, L. B., Wolt, T. B., and Leslie, J. G. (1986). Quantitation of elastin through measurement of its pentapeptide content. *Biochem. Biophys. Res. Comm.* **136**, 672–678.

Schomig, A., Neumann, F. J., Kastrati, A., Schühlen, H., Blasini, R., Hadamitzky, M., Walter, H., Zitzmann-Roth, E. M., Richardt, G., Alt, E., Schmitt, C., and Ulm, K. (1996). A randomized comparison of antiplatelet and anticoagulant therapy after the placement of coronary-artery stents. *N. Engl. J. Med.* **334**, 1084–1089.

Seeger, J. M., and Klingman, N. (1988). Improved in vivo endothelialization of prosthetic grafts by surface modification with fibronectin. *J. Vasc. Surg.* **8**, 476–482.

Sell, S. A., McClure, M. J., Barnes, C. P., Knapp, D. C., Walpoth, B. H., Simpson, D. G., and Bowlin, G. L. (2006). Electrospun polydioxanone-elastin blends: Potential for bioresorbable vascular grafts. *Biomed. Mater.* **1**, 72–80.

Smith, M. J., McClure, M. J., Sell, S. A., Barnes, C. P., Walpoth, B. H., Simpson, D. G., and Bowlin, G. L. (2008). Suture-reinforced electrospun polydioxanone-elastin small-diameter tubes for use in vascular tissue engineering: A feasibility study. *Acta Biomater.* **4**, 58–66.

Starcher, B. C., Saccomani, G., and Urry, D. W. (1973). Coacervation and ion-binding studies on aortic elastin. *Biochim. Biophys. Acta.* **310**, 481–486.

Stitzel, J., Liu, J., Lee, S. J., Komura, M., Berry, J., Soker, S., Lim, G., Van Dyke, M., Czerw, R., Yoo, J. J., and Atala, A. (2006). Controlled fabrication of a biological vascular substitute. *Biomaterials* **27**, 1088–1094.

Stone, P. J., Morris, S. M., Griffin, S., Mithieux, S., and Weiss, A. S. (2001). Building Elastin. Incorporation of recombinant human tropoelastin into extracellular matrices using nonelastogenic rat-1 fibroblasts as a source for lysyl oxidase. *Am. J. Respir. Cell Mol. Biol.* **24**, 733–739.

Swee, M. H., Parks, W. C., and Pierce, R. A. (1995). Developmental regulation of elastin production. Expression of tropoelastin pre-mRNA persists after down-regulation of steady-state mRNA levels. *J. Biol. Chem.* **270**, 14899–14906.

Toonkool, P., Jensen, S. A., Maxwell, A. L., and Weiss, A. S. (2001). Hydrophobic domains of human tropoelastin interact in a context-dependant manner. *J. Biol. Chem.* **276**, 44575–44580.

Trabbic-Carlson, K., Setton, L. A., and Chilkoti, A. (2003). Swelling and mechanical behaviors of chemically cross-linked hydrogels of elastin-like polypeptides. *Biomacromolecules* **4**, 572–580.

Urry, D. W., and Long, M. M. (1977). On the conformation, coacervation and function of polymeric models of elastin. *Adv. Exp. Med. Biol.* **79**, 685–714.

Vieth, S., Bellingham, C. M., Keeley, F. W., Hodge, S. M., and Rousseau, D. (2007). Microstructural and tensile properties of elastin-based polypeptides crosslinked with genipin and pyrroloquinoline quinone. *Biopolymers* **85**, 199–206.

Vrhovski, B., and Weiss, A. S. (1998). Biochemistry of tropoelastin. *Eur. J. Biochem.* **258**, 1–18.

Welsh, E. R., and Tirrell, D. A. (2000). Engineering the extracellular matrix: A novel approach to polymeric biomaterials. I. Control of the physical properties of artificial protein matrices designed to support adhesion of vascular endothelial cells. *Biomacromolecules* **1**, 23–30.

Wise, S. G., Mithieux, S. M., Raftery, M. J., and Weiss, A. S. (2005). Specificity in the coacervation of tropoelastin: Solvent exposed lysines. *J. Struct. Biol.* **149**, 273–281.

Wise, S. G., and Weiss, A. S. (2009). Tropoelastin. *Int. J. Biochem. Cell Biol.* **41**, 494–497.

Woodhouse, K. A., Klement, P., Chen, V., Gorbet, M. B., Keeley, F. W., Stahl, R., Fromstein, J. D., and Bellingham, C. M. (2004). Investigation of recombinant human elastin polypeptides as non-thrombogenic coatings. *Biomaterials* **25**, 4543–4553.

Wu, K. K., and Thiagarajan, P. (1996). Role of endothelium in thrombosis and hemostasis. *Annu. Rev. Med.* **47**, 315–331.

Yin, Y., Bilek, M. M., Mckenzie, D. R., Nosworthy, N. J., Kondyurin, A., Youssef, H. Byrom, M. J., and Yang, W. (2009a). Acetylene plasma polymerized surfaces for covalent immobilization of dense bioactive protein monolayers. *Surf. Coatings Technol.* **203**, 1310–1316.

Yin, Y., Wise, S. G., Nosworthy, N. J., Waterhouse, A., Bax, D. V., Youssef, H., Byrom, M. J., Bilek, M. M., McKenzie, D. R., Weiss, A. S., and Ng, M. K. (2009b). Covalent immobilisation of tropoelastin on a plasma deposited interface for enhancement of endothelialisation on metal surfaces. *Biomaterials* **30**, 1675–1681.

Zeugolis, D. I., Khew, S. T., Yew, E. S., Ekaputra, A. K., Tong, Y. W., Yung, L. Y., Hutmacher, D. W., Sheppard, C., and Raghunath, M. (2008). Electro-spinning of pure collagen nano-fibres — just an expensive way to make gelatin? *Biomaterials* **29**, 2293–2305.

THE ARCHITECTURE OF THE CORNEA AND STRUCTURAL BASIS OF ITS TRANSPARENCY

By CARLO KNUPP, CHRISTIAN PINALI, PHILIP N. LEWIS, GERAINT J. PARFITT, ROBERT D. YOUNG, KEITH M. MEEK, AND ANDREW J. QUANTOCK

Structural Biophysics Group, School of Optometry & Vision Sciences, Cardiff University, Maindy Road, Cardiff CF24 4LU, UK

Abstract

The cornea is the transparent connective tissue window at the front of the eye. In the extracellular matrix of the corneal stroma, hybrid type I/V collagen fibrils are remarkably uniform in diameter at approximately 30 nm and are regularly arranged into a pseudolattice. Fibrils are believed to be kept at defined distances by the influence of proteoglycans. Light entering the cornea is scattered by the collagen fibrils, but their spatial distribution is such that the scattered light interferes destructively in all directions except from the forward direction. In this way, light travels forward through the cornea to reach the retina. In this chapter, we will review the macromolecular components of the corneal stroma, the way they are organized into a stacked lamellar array, and how this organization guarantees corneal transparency.

I. Introduction

The cornea is the tough and transparent connective tissue at the front of the eye. Not only is it transparent, but by virtue of its convex topography it is also the main refractive element of the visual system, accounting for about 75% of the focusing power of the eye, with the lens responsible for the remainder. The cornea's high collagen content makes it physically robust, and along with the adjacent white sclera it forms the outer coat of the eye and protects against deformations, external stresses, and injuries. The properties of the cornea are a direct consequence of its complex, hierarchical molecular architecture and coassociations. In this chapter, we will provide an overview of corneal structure and describe the main molecular components of the tissue and the way they are structurally organized and distributed, touching upon how this organization gives rise to the most remarkable of the corneal characteristics — transparency.

II.   THE CORNEA — A MACROSCOPIC OVERVIEW

The cornea measures about 11 mm vertically and 12 mm horizontally in adult humans. The central area that lies directly in front of the pupil is the main optical zone and is about 3–4 mm in diameter. This prepupillary zone is the thinnest part of the cornea at about 0.5 mm thick in humans (Prakash *et al.*, 2009). Around the central area, vision researchers often refer to the midperipheral area, which is an annulus that extends from the edge of the prepupillary zone to a diameter of about 6 mm. The cornea at this point can be up to 0.7 mm thick, displaying less curvature than the central area, and is involved with peripheral vision. More radially is the peripheral cornea, and at the cornea's edge is the interface with the white sclera which, anatomically, is known as the limbus.

Under the light microscope, five different layers are recognizable through the depth of the cornea (Fig 1). The outermost is the multi-layered corneal epithelium that is approximately 50–60 μm thick in humans. Through life there is a constant shedding of superficial epithelial cells from the surface of the epithelium and a radial replenishment at the basal layer from the limbal epithelium, a process which, if disturbed, can lead to opacification and visual impairment (Kinoshita *et al.*, 2001). Immediately beneath the basement membrane of the corneal epithelium is Bowman's layer. This is 8–12 μm thick in the adult human and is a dense acellular network of collagen fibrils, which are mainly types I, III and V, with type VII anchoring filaments aiding epithelial adhesion

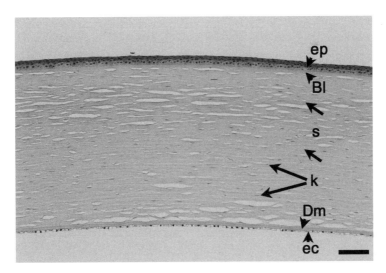

Fig. 1. Histological section of human cornea stained with hematoxylin and eosin. The cornea is surfaced by a multilayered epithelium (ep), overlying Bowman's layer (Bl), a specialized outer layer of the stroma(s), which represents 90% of the full tissue thickness. The bulk of the stroma comprises a stacked array of collagenous layers or lamellae (arrows) with scattered cells, termed keratocytes (k). Some separation of lamellae, appearing here as gaps in the stroma, occurs during tissue processing. Descemet's membrane (Dm) forms the innermost boundary of the stroma and supports a monolayer of endothelial cells (ec) that line the anterior chamber. Bar = 1 mm. Image courtesy of Dr. Hidetoshi Tanioka, Department of Ophthalmology, Kyoto Prefectural University of Medicine, Japan.

(Komai and Ushiki, 1991; Wilson and Hong, 2000) (Fig. 2). Collagen fibrils in Bowman's layer are thinner than those in the stroma proper and numerous lamellar insertions exist which integrate Bowman's layer and the subjacent stroma (Morishige et al., 2007). The role of Bowman's layer is not entirely clear, but it has been hypothesized to act to maintain corneal structural integrity or as a barrier against viral penetration (Wilson and Hong, 2000). Most of the cornea's thickness is made up by the stroma, a stacked lamellar array of collagen fibrils of constant diameter and uniform spatial organization which are interspersed by keratocytes: the corneal fibroblasts. The human stroma has over 200 lamellae (Fig. 3), which are highly interwoven in the anterior stroma, but thicker and more stacked and distinct in the deeper tissue (Hogan et al., 1971; Komai and

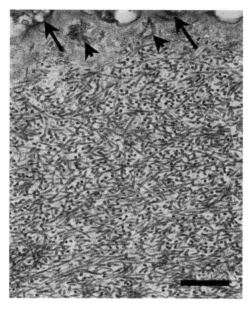

Fɪɢ. 2. Electron micrograph showing nonlamellar, meshwork arrangement of collagen fibrils in Bowman's layer of corneal stroma in human eye. Anchoring fibrils (arrowheads), composed of type VII collagen, attach Bowman's layer to a type IV collagen-rich basal lamina, seen here as a diffuse amorphous structure, underlying the proximal membranes of basal epithelial cells (arrows). Bar = 500 nm.

Ushiki, 1991). Within each lamella, collagen fibrils tend to be aligned in the same direction and are surrounded by small leucine-rich protein–carbohydrate molecules called proteoglycans. The light transmission and mechanical properties of the cornea are defined to a large extent by the stroma, and this will be discussed in more detail later. Adhered to the posterior-most stroma is the thin Descemet's membrane, the basement membrane of the corneal endothelium that progressively thickens with age (Beuerman and Pedroza, 1996). Immunohistochemical studies have found that fibronectin, type IV collagen, and laminin are present in Descemet's membrane (Beuerman and Pedroza, 1996), as is type VIII collagen that forms a hexagonal lattice (Shuttleworth, 1997). The innermost corneal layer is the endothelium. It is a monolayer, 4–6 μm thick, and formed of approximately 500,000 cells (Hogan *et al.*, 1971). Corneal endothelial cells do

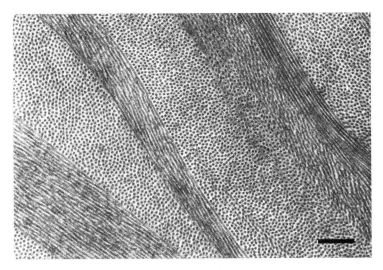

FIG. 3. Ultrastructure of mid stroma in monkey cornea showing architecture of stacked lamellae within each of which collagen fibrils exhibit uniform orientation, diameter, and spacing. Bar = 500 nm.

not regenerate under normal conditions in adult humans, and the layer itself has functional reserve due to high cell numbers. The endothelial cells play a fundamental role in controlling the hydration level and swelling tendencies of the corneal stroma (Hodson and Miller, 1976; Maurice, 1972).

This chapter will focus on the structure and function of the corneal stroma. To a large extent, corneal properties are defined by the properties of collagens and proteoglycans, and these are discussed below.

III. COLLAGEN

Collagen is the most abundant protein in connective tissues. The function of collagen has always been considered mainly a structural one, but the collagen family, made up of several genetically distinct types, carries out many other different functions. For example, collagens are also involved in cell attachment and differentiation, as chemotactic agents, and in immunopathological processes as antigens (Linsenmayer, 1991).

The common structural feature of all collagens is that they possess one or more triple helical domains. These domains consist of three polypeptide chains (called α-chains) wound around a common axis in a right-handed fashion. Every individual chain is made up of a sequence of amino acids forming a left-handed coil. This sequence is characterized by the repeating motif G-X-Y, where G represents glycine and X and Y can be any amino acids, although they often are proline and hydroxyproline, respectively. In the triple helix, every α-chain is staggered axially by one residue with respect to the other two α-chains, so that all glycines, the smallest of the amino acids, can be accommodated internally in the central axis of the triple helix. Interchain hydrogen bonds between the amino group of the glycine and the carbonyl group of the amino acid in the X position stabilize the collagen triple helix. The triple helix is such that all peptide bonds linking adjacent amino acids are buried within the molecule, making it resistant to general proteases such as pepsin (Bella *et al.*, 1994; Linsenmayer, 1991; Okuyama, 2008; Rich and Crick, 1961). The side chains of the residues in the X and Y positions face the external environment, mediating all interactions with cells and other molecules.

The roles and tissue distribution of collagen are varied. At least twenty-nine genetically distinct types of collagen exist and they can be subdivided into families such as fibril-forming collagens, fibril-associated collagens with interrupted triple helix (FACIT), network-forming collagens, anchoring fibril collagens, transmembrane collagens, and multiplexins (Gelse *et al.*, 2003; Knupp and Squire, 2005; Prockop and Kivirikko, 1995; van der Rest and Garrone, 1991). The first type of collagen to be studied (type I collagen) is also the most abundant and is found in the cornea, skin, bone, tendon as well as in several other tissues. Over the years, it has become apparent that there are many different types of collagen. To each of them a Roman numeral was assigned (I, II, III, IV, etc.) according to the chronological order in which the particular collagen was discovered.

About 70% of the dry weight of a cornea is from collagen, most of which is fibrillar. Collagen types I, II, III, V, VI, XII, and XIV have been found in the corneal stroma, with type I being predominant (Marshall *et al.*, 1991, 1993; Newsome *et al.*, 1982). Of the collagens in the cornea, types I, II, III, and V belong to the fibril-forming collagen family and are similar to each other in size and their possession of an uninterrupted triple helical domain containing about 340 G-X-Y triplets per α-chain. The most striking structural feature of these fibril-forming collagens is the staggered manner in which the molecules assemble. As a result, they give rise to fibrils with variable electron-dense regions and a 67-nm axial periodicity in most tissues. When stained with uranyl acetate before examination in a transmission electron microscope, a characteristic banding pattern appears within the fibril's 67-nm repeat, and for ease of description these bands are labeled sequentially from "a" to "e" (Chapman *et al.*, 1990).

Type I collagen is the most abundant of all collagens, and it is usually found in heterotrimeric form with two $\alpha 1$ and one $\alpha 2$ chains. In the cornea, the axial periodicity of the fibrils is reduced to 65 nm (Meek *et al.*, 1981), possibly because of a tilt of the collagen molecules with respect to the fibril axis (Holmes *et al.*, 2001) (Fig. 4). Type II collagen is a homotrimer and is found in the developing cornea of the eye (Gelse *et al.*, 2003). Type III collagen is also a homotrimer and is found in human cornea during wound healing (Nakamura, 2003). Type V collagen is one of the most heterogeneous collagens because its three α-chains belong to at least four distinct genetic types. Moreover, it can also contain a type XI collagen α-chain (Ricard-Blum and Ruggiero, 2005). Type V collagen can associate with type I (Birk, 2001) and III collagens (Gelse *et al.*, 2003). In the corneal stroma type V collagen coassembles with type I collagen (Birk *et al.*, 1988). A feature of type V collagen is that it retains its N-terminal domain after cleavage (Birk, 2001; Wu and Eyre, 1995), and *in vitro* experiments suggest that the interaction of types I and V collagens plays a role in the regulation of the diameter of the fibrils formed in the cornea (Birk *et al.*, 1990). It is possible that this occurs because of the retained N-terminus, which may

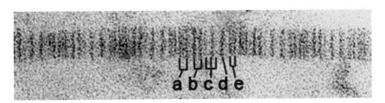

FIG. 4. Electron micrograph of longitudinally sectioned collagen fibril in bovine corneal stroma. The "molecular staining pattern" of collagen — a series of periodic bands, "a–e," is revealed by positive staining with uranyl acetate and represents the distribution of charged amino acids along the fibril (Chapman, 1974). Altogether, twelve bands are present, although the four "a" bands and outer "c" (c_1 and c_3) bands are difficult to distinguish in sections of intact tissue.

limit the lateral accretion of the heterotypic type I/V fibrils (Adachi and Hayashi, 1985; Kadler *et al.*, 1996; Marshall *et al.*, 1993; Wess, 2005), although other molecular mechanisms have been suggested for this fibril diameter regulatory property (Scott and Parry, 1992) including the influence of stroma proteoglycans (Rada *et al.*, 1993).

Type VI collagen is widespread throughout the human body and is found in the cornea (Eyre, 1991; Poole *et al.*, 1992; Reale *et al.*, 2001; von der Mark *et al.*, 1984), but is not thought to participate in fibril formation (Marshall *et al.*, 1993). It is a rod-like molecule whose triple helical portion is about 105 nm long (von der Mark *et al.*, 1984). Type VI collagen exploits disulfide bonds to form antiparallel supercoiled dimers that are axially staggered by about 30 nm (Knupp and Squire, 2001). These dimers assemble both linearly, to form beaded filaments, and laterally via their bulky C- and N-terminal domains, to form open networks (Knupp and Squire, 2005). In the eye, such networks are found in association with diseases such as age-related macular degeneration and Sorsby's fundus dystrophy (Knupp *et al.*, 2002a,b, 2006). In the cornea, type VI collagen can be found in between the fibrils as beaded filaments (Marshall *et al.*, 1991) and, on occasion, as open networks.

Type XII and XIV collagens colocalize with type I collagen in the skin and other tissues. They belong to the FACIT group of nonfibrillar collagens comprising triple helical domains that are interrupted with non-collagenous domains. Normally, these collagens are linked to the surface of a host fibrillar collagen via the collagenous domain at one extremity of the molecule, with the rest of the collagen projected outward toward the external environment. A noncollagenous domain acts as a hinge to achieve this, and this conformation is thought to facilitate interactions with other extrafibrillar matrix proteins or cells (Gelse *et al.*, 2003; van der Rest and Garrone, 1991). In the developing cornea, type XII collagen may have a role in maintaining fibril organization, whereas type XIV collagen may regulate fibrillogenesis (Young *et al.*, 2002).

IV. GLYCOSAMINOGLYCANS AND PROTEOGLYCANS

Structurally, glycosaminoglycans are unbranched chains of repeating disaccharide units in which one of the monosaccharides is an amino sugar, and one or both monosaccharides contain a sulfate or carboxylate group (Gandhi and Mancera, 2008; Garrett and Grisham, 2008; Kjellén and Lindahl, 1991;). At physiological pH, the sulfate and carboxylate groups are charged, providing glycosaminoglycans with a very high negative charge density (Gandhi and Mancera, 2008). The most common glycosaminoglycans are hyaluronate (a macromolecule that does not participate directly in the formation of proteoglycans by means of covalent bonds, but can interact with them), chondroitin sulfate, dermatan sulfate, keratan sulfate, heparin, and heparan sulfate (Gandhi and Mancera, 2008; Garrett and Grisham, 2008; Kjellén and Lindahl, 1991). Hyaluronate is usually found in synovial fluid, vitreous humor, and the extracellular matrix of loose connective tissues. Chondroitin sulfate is prevalent in the cartilage, tendon, ligament, and aorta, while dermatan sulfate is found in the skin, blood and heart vessels, and keratan sulfate in the cartilage. Versican, also a chondroitin sulfate-based proteoglycan, has been found in chick and rat embryo corneas, and it is suggested to form large molecular complexes with hyaluronan, also present in embryonic corneas, to play a role in corneal development (Koga et al., 2005).

Proteoglycans in the cornea possess small globular protein cores, to which one or more glycosaminoglycan chains are covalently attached. Some investigators have reported that the core proteins of decorin, which folds as a curved solenoid, crystallize as antiparallel dimers. The two monomers thus have their concave faces interacting with each other, suggesting that these surfaces would not be available to form bonds with collagen molecules (Scott et al., 2004). However, this has been refuted by other groups who claim that biologically active decorin is a monomer in solution (Goldoni et al., 2004).

Complexes made up of one or more glycosaminoglycan chains can covalently link to a protein core to form a molecule known as a proteoglycan. Proteoglycans have been implicated in cell adhesion mechanisms, cell motility, proliferation, differentiation, and morphogenesis. These functions are often due to the interaction of proteins with the glycosaminoglycan chains, which can have low affinity and specificity and depend on charge interactions, or can have high affinity and specificity and involve regions of the polysaccharide chain with conserved structure (Kjellén and Lindahl, 1991). Because of the negative charge of the sulfate groups in the glycosaminoglycan chains, proteoglycans can make ionic interactions with conveniently located arginines, lysines, and histidines on the surface of proteins. Strong ionic bonds are also made with divalent metal ions (Ca^{2+}, Mg^{2+}, Mn^{2+}) in proteins or in solution (Gandhi and Mancera, 2008). Glycosaminoglycan chains can also interact with proteins by means of van der Waals forces, hydrogen forces, and hydrophobic forces. The glycosaminoglycans are thought to assume extended conformations in water so that their sulfated groups can maximize their hydrophilic interactions with water molecules (Imberty *et al.*, 2007).

In the extracellular matrix of connective tissues, proteoglycans play a major role in opposing compressive mechanical forces. Negatively charged glycosaminoglycans tend to swell in solution to increase their conformational entropy. Mutual repulsion of the glycosaminoglycan charges and Donnan osmotic pressure, which is a consequence of the positively charged ions gravitating around the glycosaminoglycan chains, also contribute to their swelling (Scott, 2003). Proteoglycans vary considerably in structure because of the differential expression of the genes encoding the core proteins and because of variations in length and type of the glycosaminoglycan chains and in the position of the chain's sulfated groups (Gandhi and Mancera, 2008; Imberty *et al.*, 2007). The sulfation status of corneal proteoglycans is based on enzymatic activity (Akama *et al.*, 2001; Hayashida *et al.*, 2006), is changed with development (Quantock and Young, 2008), after injury (Hassell *et al.*, 1983), and is altered in some inherited diseases (Akama *et al.*, 2000; Hassell *et al.*, 1980; Quantock *et al.*, 1990). In the extracellular matrix, proteoglycans play an important role in providing mechanical support, functioning as shock absorbers and modulating collagen fibrillogenesis (Kjellén and Lindahl, 1991).

In the cornea, the main glycosaminoglycans are keratan sulfate and a chondroitin sulfate/dermatan sulfate hybrid (Fig. 5). These are bound to various core proteins to form proteoglycans, with lumican, keratocan, and mimecan bearing keratan sulfate glycosaminoglycan chains, and decorin and biglycan possessing chondroitin sulfate/dermatan sulfate side chains (Blochberger *et al.*, 1992; Corpuz *et al.*, 1996; Funderburgh *et al.*, 1997, 1998; Koga *et al.*, 2005; Li *et al.*, 1992; Liu *et al.*, 1998). *In vitro* experiments of collagen fibrillogenesis have shown that the core protein of decorin, with or without its glycosaminoglycan side chain(s), can restrict the diameter to which collagen fibrils grow, and the lumican proteoglycan can act in a similar manner (Rada *et al.*, 1993; Raspanti *et al.*, 2008). The proteoglycans have a role *in situ*, too, with the corneas of lumican-null mice [in which keratocan is also affected (Carlson *et al.*, 2005)] and compound decorin/biglycan mutant mice, possessing pockets

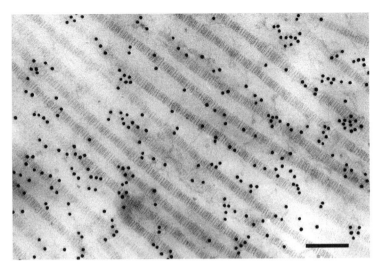

FIG. 5. Electron micrograph of bovine cornea stroma swollen in 0.15 M saline. Collagen fibrils in longitudinal section are decorated with 10 nm gold particles indicating monoclonal antibody 5D4 bound to sites of highly sulfated glycosaminoglycans on keratan sulfate proteoglycans. Bar = 150 nm.

of larger-than-normal collagen fibrils with irregular profiles (Chakravarti et al., 1998, 2000; Zhang et al., 2009).

A comparative electron histochemical study has shown that proteoglycans generally associate with collagen fibrils at the a, c, d, and e bands within the 65-nm D-period and that keratan sulfate proteoglycans occupy the a and c bands, with chondroitin/dermatan sulfate proteoglycans at the d/e bands in the gap zone (Scott and Haigh, 1985). Keratan sulfate is more abundant in the corneas of animals which have thick corneas, whereas the amount of chondroitin and dermatan sulfate is lower (Scott and Bosworth, 1990). Thus, the total content of sulfated glycosaminoglycan is constant across species. As first suggested by Scott and Haigh (1988), this may be a consequence of the fact that the synthesis of keratan sulfate requires no oxygen, while chondroitin sulfate is synthesized in the presence of oxygen. Possibly, the longer oxygen diffusion paths in thicker corneas of larger animals increase the probability of oxygen being consumed before reaching all regions and may therefore determine a preference for the synthesis of keratan sulfate in thicker corneas, especially in deeper stromal regions. The even amount of sulfated proteoglycan across species could explain the similar level of hydration — at about 80% — and similar interfibrillar volume between species (Meek and Leonard, 1993), if the ability of attracting water were the same for the sulfated chains of chondroitin, dermatan, and keratan sulfate proteoglycans. The concept put forward by Scott is that thin corneas rely mostly on chondroitin/dermatan sulfate with thicker corneas having additional input from keratan sulfate (Scott and Bosworth, 1990).

The length and thickness of chondroitin/dermatan sulfate glycosaminoglycan chains isolated from cow and rabbit corneas, when measured by electron microscopy after staining with cupromeronic blue, was found to be greater than that of keratan sulfate proteoglycans (Scott, 1992a). This was true for stained proteoglycans measured in intact tissue sections of rabbit, rat, and mouse corneas too. Stained glycosaminoglycans in tissue sections sometimes appeared twice as long as their isolated counterparts, and based on this, Scott proposed the presence of two glycosaminoglycan chains, 180° apart, linked to a single proteoglycan core protein. In the electron microscope, it would not be possible to distinguish the two different chains because they would appear as one long entity. Glycosaminoglycan chains are seen to extend between neighboring collagen fibrils, and it was hypothesized that the chains could form antiparallel associations, with the proteoglycan protein cores bound to neighboring collagen fibrils and the glycosaminoglycan chains joined together to form a bridge between the fibrils. Hydrophobic and hydrogen bonding could potentially stabilize antiparallel associations (Scott, 2001). In this manner, Scott highlighted the possibility that the collagen interfibrillar spacing is controlled by proteoglycans (1992b). Because of the high sulfation of the glycosaminoglycan chains, which results in a high negative charge density, it is difficult to envisage how two chains can form durable antiparallel associations without being disrupted by repulsive forces. However, because of the variations in the sulfation pattern along the glycosaminoglycan chains, a situation could arise in which negative charges are systematically missing on one or other of the interacting chains, so that two negative charges are never in close proximity along the fully extended antiparallel chains. Alternatively, it may be that such charges are screened by positive ions present in the cornea while the antiparallel associations take place. It is interesting to speculate as to whether or not the interaction between the glycosaminoglycan chains could be that of an antiparallel double helix. Indeed, this configuration is assumed by hyaluronate as seen in X-ray fiber diffraction experiments (Arnott *et al.*, 1983; Dea *et al.*, 1973). The hyaluronate backbone is only slightly different from that of the corneal glycosaminoglycans, making it plausible that chondroitin/dermatan sulfate and keratan sulfate glycosaminoglycans behave in a similar way.

V. THE STRUCTURE OF THE CORNEAL STROMA

Aside from Bowman's layer, all collagen fibrils in the cornea are organized into layers which are referred to as lamellae. Within each lamella, fibril axes generally run in the same direction and are parallel to the corneal surface (Komai and Ushiki, 1991). The fibril direction in adjacent lamellae, however, varies. In humans, the thickness of an individual lamella is between 0.2 and 2.5 μm, being thinner and more interwoven in the anterior one-third of the stroma and thicker and less interwoven posteriorly (Komai and Ushiki, 1991). X-ray diffraction experiments on the central prepupillary zone of the human cornea indicate two preferred orientations for the lamellae and their component fibrils in the plane of the cornea. These are 90° apart and in temporal–nasal and superior–inferior orientations (Boote *et al.*, 2005; Meek *et al.*, 1987). However, this structural arrangement does not occur in the corneas of all species (Hayes *et al.*, 2007), and is thought to be linked to the muscle insertion points and biomechanical stability in the human cornea (Boote *et al.*, 2005; Meek and Boote, 2009). Recently, it was discovered that this arrangement exists primarily in the posterior two-thirds of the stroma, with the lamellae in the anterior one-third of the cornea being less preferentially aligned (Abahussin *et al.*, 2009). More peripherally in the human cornea, collagen fibrils form a distinct annulus at the limbus (Aghamohammadzadeh *et al.*, 2004; Newton and Meek, 1998). In this region, some of the orthogonal collagens are thought to change direction and merge with circumferential fibers. Some fibrils from the sclera could also extend to the peripheral regions of the cornea and contribute to the structural integrity of the limbus (Meek and Boote, 2009).

Collagen fibril diameters and interfibrillar distances are remarkably uniform in the central cornea, and this is an essential prerequisite for corneal transparency. In humans, collagen fibril diameter in the hydrated stroma is about 31 nm, although this can increase slightly with age (Daxer *et al.*, 1998; Meek *et al.*, 1989). Fibril diameter is constant across the cornea in humans, whereas the average interfibrillar spacing is 5–7% less in more peripheral corneal regions when compared to the prepupillary zone (Boote *et al.*, 2003). Interestingly, aquatic animals tend to have smaller fibril diameters and smaller interfibrillar spacing than terrestrial animals. A constant feature comparatively is the volume of the stroma occupied by fibrils as a ratio of the extrafibrillar space, known as the fibril volume fraction (Meek and Leonard, 1993).

Several models to explain how collagen fibrils are kept in position in the corneal stroma have been proposed (Farrell and Hart, 1969; Farrell *et al.*, 1973; Maurice, 1962; Müller *et al.*, 2004). Given that proteoglycans occupy the extrafibrillar space (Fig. 6), many of these models are primarily based on the interactions between proteoglycans and collagen fibrils. In the late 1960s, Farrell and Hart (1969) considered possible models in which proteoglycans' glycosaminoglycan chains formed bridges between adjacent collagen fibrils. The simplest of these models featured six

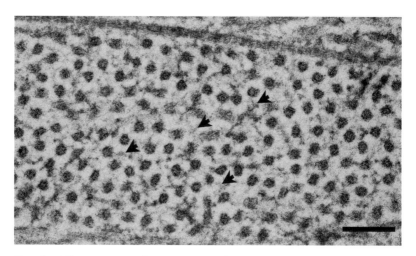

Fig. 6. Ultrastructure of mouse corneal stroma prepared by high pressure cryofixation and freeze substitution in acetone with osmium tetroxide to better preserve native hydrated structure. A lamella with collagen fibrils in transverse section reveals interfibrillar structures, presumed to be proteoglycans (arrowheads), preserved in the absence of cationic stain. Bar = 150 nm.

glycosaminoglycan chains radiating from a collagen fibril and connecting to six adjacent fibrils. More complex models were also proposed, in which the proteoglycan protein core did not interact directly with collagen, but was positioned away from the fibrils and acted as a mediator between the glycosaminoglycan chains that connected the fibrils. All of these models were based on a systematic arrangement of the proteoglycans around the collagen fibrils that implied sixfold symmetry for the collagen–proteoglycan arrangement. More recently, Müller and collaborators (2004) put forward a modified version of Farrell and Hart's simplest model that also featured the proteoglycan bridges between collagen fibrils. Muller's model was based on transmission electron microscopy data, where glyco-saminoglycan chains connecting three adjacent fibrils in longitudinal views were observed. The proposed model featured six proteoglycan bridges radiating from a collagen fibril and binding to six other fibrils. These were not the nearest available fibrils, but the next nearest neighbors. Axially, these bridges labeled the fibrils in a periodic fashion, and Muller's model implies sixfold hexagonal symmetry for the collagen–proteoglycan arrangement. By extension to the whole stroma, these models would predict a near perfect sixfold arrangement for the collagen fibrils. Careful scrutiny of transverse views from the corneal stroma from several species suggests that the collagen fibril arrangement does not show a perfect sixfold symmetry, even allowing for disruptions caused by sample preparation techniques. In addition, precise sixfold symmetry is not required for corneal transparency, just some degree of regularity in the collagen arrangement (this will be discussed in more depth in Section VI). To try to reconcile the existing models with the overall appearance of the corneal stroma, we recently obtained three-dimensional recon-structions of cow and mouse corneal stroma by electron tomography of cupromeronic blue-stained, resin-embedded tissue. The data show that the interactions between collagen fibrils and stained proteoglycan fila-ments are not simple and sixfold, but more complex than previously thought (Fig. 7). The model proposed to explain that our three-dimensional data has proteoglycans that attach to the collagen fibrils by their protein cores, but with no predefined azimuthal position. These proteoglycans can interact via their glycosaminoglycan chains with other proteoglycans on nearby fibrils forming antiparallel multiplexes. Although there is no systematic sixfold arrangement of proteoglycans around any collagen fibrils, there are bridges that connect all adjacent

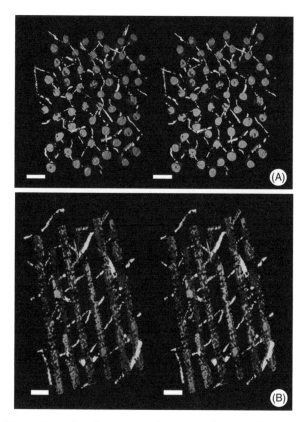

Fɪɢ. 7.   Three-dimensional reconstruction from the central anterior region of mouse cornea. (A) Transverse view. (B) Longitudinal view. Collagen fibrils are painted in blue and proteoglycans in yellow. The proteoglycans shown in the reconstructions are mainly those with chondroitin/dermatan sulfate glycosaminoglycan chains. No apparent regular proteoglycan labeling of the collagen fibrils is visible. Bars = 50 nm. (See Color Insert.)

fibrils at different axial positions, with the overall effect that a pseudo-sixfold arrangement of fibrils arises. An extension of this arrangement to the whole corneal stroma would explain better what is seen by electron microscopy.

Our tomographic electron microscopy observations also led us to propose a mechanism to explain how the distance between adjacent fibrils is maintained. This mechanism is based on the opposing effects of two forces: a repulsive force trying to push the fibrils apart and an attractive force trying to bring the fibrils together. The repulsive force would arise because of an accumulation of water molecules in the volume between the fibrils that increases local pressure. Water molecules would migrate between fibrils by osmosis and be ultimately driven there by the presence of the charged glycosaminoglycan chains (the Donnan effect). The attractive forces would arise by thermal motion of the glycosaminoglycan chains themselves. In fact, the bridge-forming proteoglycan complexes can be thought of as being constantly vibrating, so that their glycosaminoglycan chains deviate from a fully extended configuration. The distance between the terminal ends of the bridges decreases by an amount proportional to the amplitude and the number of vibrational modes of the glycosaminoglycan chains. Because the terminal ends are attached to collagen fibrils via the proteoglycan core protein and are pulled toward each other by chain vibrations, an attractive force arises between the fibrils. Since both proposed attractive and repulsive forces are thermal in nature and increase in strength with temperature, they may also act to keep the interfibrillar distances constant over a range of temperatures. The picture of the cornea arising from this model is a dynamic one, where all macromolecules are in constant relative motion. This may facilitate processes such as the transport of oxygen and nutrients through the cornea as well as cell migration.

VI. Corneal Transparency

The most convincing explanation for corneal transparency was put forward by Maurice in the 1950s (1957). In his explanation, visible light with defined frequencies would enter the cornea from the outside world. Here it interacts with the collagen fibrils in the stroma. A small portion of the energy of the incoming light is transferred to the electrons in the collagen fibrils, and these would start to oscillate with a frequency identical to that of the incoming electromagnetic waves. Due to their oscillatory movement, the electrons become a source of secondary radiation, so that all collagen fibrils themselves radiate light in all directions. A collagen fibril can be thought of as a very long cylinder with small radius because it is generally much longer than the areas affected by the incoming waves. Thus, it is possible to simplify the expression describing the secondary radiation from any fibril, which is dependent on the wavelength of the incoming light, the collagen fibril diameter, and the refractive indices of the fibril and of the ground substance between fibrils. Even though the secondary light from each collagen fibril is a small portion of the incoming radiation, the fact that there are so many fibrils in the corneal stroma signifies that almost all of the incoming light would be radiated in all directions. A consequence of this is that corneal transparency would be greatly impaired since very little light can propagate in the forward direction. But Maurice noticed that the positions of the collagen fibrils in the cornea are far from being random. In fact, to a first approximation, he noticed that the fibrils can be thought of as sitting on a hexagonal lattice and considered the consequences of this observation. If there are strict positional constraints for the collagen fibrils, it is possible for the secondary radiation emanating from them to interfere, leading to constructive and/or destructive interference. A relatively simple calculation shows that if the diameter is the same for all fibrils, and if the fibrils sit on a hexagonal lattice whose sides are smaller than half the wavelength of the incoming light, crests will sum to troughs in all directions except the forward direction. In other words, all secondary light will cancel out in all directions and will only be able to propagate in the forward direction, thus making the cornea transparent. Successive theoretical investigations following on from Maurice highlighted the fact that the collagen fibrils do not need to form a perfect hexagonal lattice (Cox *et al.*, 1970; Hart and Farrell, 1969), but that, aside from uniformity in fibril diameter, it is the distribution of distances between adjacent fibrils that must be very narrow (i.e., the distance between adjacent fibrils must be more or less the same but still less than half the incoming wavelengths). This, in fact, is an accurate description of the positional constraints obeyed by the collagen fibrils in the cornea, which is likely to be a direct consequence of the way fibrils are kept in position by the proteoglycans as discussed above.[1] An upshot is that if either diseases or adverse physiological conditions impair the uniformity of the distances between adjacent fibrils, or the uniformity of the fibril diameters, corneas can lose their transparency. Other

---

[1] A very readable technical account of corneal transparency theory was published by Benedek (1971), and papers that outline computational techniques that can be easily used to predict the transparency of corneas from electron micrographs were written by Freund and collaborators (1986, 1995). Readers are referred to these papers for mathematical details of corneal transparency theory.

physical phenomena should be taken into account when considering theories of corneal transparency. These include the possibility of multiple scattering (i.e., secondary waves from collagen fibrils that excite electrons in other fibrils that emit tertiary waves), the effect of which was examined by Smith (1988) and shown to be small, incoherent scattering (secondary waves with frequencies that are different from those of the primary waves), internal reflections between corneal lamellae, light absorption by matrix molecules, and the fact that there are cells within the cornea that are behaving differently, from an optical point of view, from the extracellular matrix.

VII. Summary

The transparency of the cornea is a direct consequence of the structural organizations of the macromolecular components of the stroma, the predominant layer in the cornea. The collagen fibrils are kept in defined positions by the proteoglycans via nonsystematic interactions. Although the collagen fibrils do not sit on a perfect crystalline lattice, the degree of order in the stroma is sufficient to guarantee constructive interference of the waves by the collagen fibrils only in the forward direction. In this way, light can travel through the cornea and reach the retina.

Acknowledgments

This work was funded by the BBSRC Research Grants BB/F022077/1 and BB/D001919/1 to CK and AJQ, and EPSRC Research Grant EP/F034970/1 to AJQ, RDY, and KMM.

References

Abahussin, M., Hayes, S., Knox Cartwright, N. E., Kamma-Lorger, C. S., Khan, Y., Marshall, J., and Meek, K. M. (2009). 3D collagen orientation study in human cornea using x-ray diffraction and femtosecond laser technology. *Invest. Ophthalmol. Vis. Sci.* **50**, June 10 (Epub ahead of print), in press.

Adachi, E., and Hayashi, T. (1985). In vitro formation of fine fibrils with a D-periodic banding pattern from type V collagen. *Coll. Relat. Res.* **5**, 225–232.

Aghamohammadzadeh, H., Newton, R. H., and Meek, K. M. (2004). X-ray scattering used to map the preferred collagen orientation in the human cornea and limbus. *Structure* **12**, 249–256.

Akama, T. O., Nakayama, J., Nishida, K., Hiraoka, N., Suzuki, M., McAuliffe, J., Hindsgaul, O., Fukuda, M., and Fukuda, M. N. (2001). Human corneal GlcNac 6-O-sulfotransferase and mouse intestinal GlcNac 6-O-sulfotransferase both produce keratan sulfate. *J. Biol. Chem.* **276**, 16271–16278.

Akama, T. O., Nishida, K., Nakayama, J., Watanabe, H., Ozaki, K., Nakamura, T., Dota, A., Kawasaki, S., Inoue, Y., Maeda, N., Yamamoto, S., Fujiwara, T., et al. (2000). Macular corneal dystrophy type I and type II are caused by distinct mutations in a new sulphotransferase gene. *Nat. Genet.* **26**, 237–241.

Arnott, S., Mitra, A. K., and Raghunathan, S. (1983). Hyaluronic acid double helix. *J. Mol. Biol.* **169**, 861–872.

Bella, J., et al. (1994). Crystal and molecular structure of a collagen-like peptide at 1.9 Å resolution. *Science* **266**, 75–81.

Benedek, G. B. (1971). Theory of transparency of the eye. *Appl. Opt.* **10**, 459–473.

Beuerman, R. W., and Pedroza, L. (1996). Ultrastructure of the human cornea. *Microsc. Res. Tech.* **33**, 320–335.

Birk, D. E. (2001). Type V collagen: Heterotypic type I/V collagen interactions in the regulation of fibril assembly. *Micron* **32**, 223–237.

Birk, D. E., et al. (1990). Collagen fibrillogenesis in vitro: Interaction of types I and V collagen regulates fibril diameter. *J. Cell Sci.* **95**, 649–657.

Birk, D. E., et al. (1988). Collagen type I and type V are present in the same fibril in the avian corneal stroma. *J. Cell Biol.* **106**, 999–1008.

Blochberger, T. C., Vergnes, J. P., Hempel, J., and Hassell, J. R. (1992). cDNA to chick lumican (corneal keratan sulfate proteoglycan) reveals homology to the small interstitial proteoglycan gene family and expression in muscle and intestine. *J. Biol. Chem.* **267**, 347–352.

Boote, C., et al. (2003). Collagen fibrils appear more closely packed in the prepupillary cornea: Optical and biomechanical implications. *Invest. Ophthalmol. Vis. Sci.* **44**, 2941–2948.

Boote, C., Dennis, S., Huang, Y., Quantock, A. J., and Meek, K. M. (2005). Lamellar orientation in human cornea in relation to mechanical properties. *J. Struct. Biol.* **149**, 1–6.

Carlson, E. C., Liu, C.-Y., Chikama, T.-I., Hayashi, Y., Kao, C. W.-C., Birk, D. E., Funderburgh, J. L., Jester, J. V., and Kao, W. W. (2005). Keratocan, a cornea-specific keratan sulphate proteoglycan, is regulated by lumican. *J. Biol. Chem.* **280**, 25541–25547.

Chakravarti, S., Magnuson, T., Lass, J. H., Jepsen, K. J., LaMantia, C., and Carroll, H. (1998). Lumican regulates collagen fibril assembly: Skin fragility and corneal opacity in the absence of lumican. *J. Cell Biol.* **141**, 1277–1286.

Chakravarti, S., Petroll, W. M., Hassell, J. R., Jester, J. V., Lass, J. H., Paul, J., and Birk, D. E. (2000). Corneal opacity in lumican-null mice: Defects in collagen fibril structure and packing in the posterior stroma. *Invest. Ophthalmol. Vis. Sci.* **41**, 3365–3373.

Chapman, J. A. (1974).The staining pattern of collagen fibrils. I. An analysis of electron micrographs. *Connect. Tissue Res.* **2**, 137–150.

Chapman, J. A., et al. (1990). The collagen fibril — A model system for studying the staining and fixation of a protein. *Electron Microsc. Rev.* **3**(1), 143–182.

Corpuz, L. M., Funderburgh, J. L., Funderburgh, M. L., Bottomley, G. S., Prakash, S., and Conrad, G. W. (1996). Molecular cloning and tissue distribution of keratocan. Bovine corneal keratan sulfate proteoglycan 37A. *J. Biol. Chem.* **271**, 9759–9763.

Cox, J. L., *et al.* (1970). The transparency of the mammalian cornea. *J. Physiol.* **210**, 601–616.

Daxer, A., *et al.* (1998). Collagen fibrils in the human corneal stroma: Structure and aging. *Invest. Ophthalmol. Vis. Sci.* **39**, 644–648.

Dea, I. C., *et al.* (1973). Hyaluronic acid: A novel, double helical molecule. *Science* **179**, 560–562.

Eyre, D. R. (1991). The collagens of articular cartilage. *Semin. Arthritis Rheum.* **21**(3 Suppl. 2), 2–11.

Farrell, R. A., and Hart, R. W. (1969). On the theory of the spatial organization of macromolecules in connective tissue. *Bull. Math. Biophys.* **31**, 727–760.

Farrell, R. A., McCally, R. L., and Tatham, P. E.R. (1973). Wave-length dependencies of light scattering in normal and cold swallen rabbit corneas and their structural implications. *J. Physiol.* **233**, 589–612.

Freund, D. E., McCally, R. L., and Farrell, R. A. (1986). Direct summation of fields for light scattering by fibrils with applications to normal corneas. *Appl. Opt.* **25**, 2739.

Freund, D. E., *et al.* (1995). Ultrastructure in anterior and posterior stroma of perfused human and rabbit corneas. Relation to transparency. *Invest. Ophthalmol. Vis. Sci.* **36**, 1508–1523.

Funderburgh, J. L., Corpuz, L. M., Roth, M. R., Funderburgh, M. L., Tasheva, E. S., and Conrad, G. W. (1997). Mimecan, the 25-kDa corneal keratan sulfate proteoglycan, is a product of the gene producing osteoglycin. *J. Biol. Chem.* **272**, 28089–28095.

Funderburgh, J. L., Hevelone, N. D., Roth, M. R., Funderburgh, M. L., Rodrigues, M. R., Nirankari, V. S., and Conrad, G. W. (1998). Decorin and biglycan of normal and pathologic human corneas. *Invest. Ophthalmol. Vis. Sci.* **39**, 1957–1964.

Gandhi, N. S., and Mancera, R. L. (2008). The structure of glycosaminoglycans and their interactions with proteins. *Chem. Biol. Drug Des.* **72**, 455–482.

Garrett, R. H., and Grisham, C. M. (2008). "Biochemistry." 4th edn. Thomson Brooks/Cole, Pacific Grove, California.

Gelse, K., Pöschl, E., and Aigner, T. (2003). Collagens — structure, function, and biosynthesis. *Adv. Drug Deliv. Rev.* **55**, 1531–1546.

Goldoni, S., Owens, R. T., McQuillan, D. J., Shriver, Z., Sasisekharan, R., Birk, D. E., Campbell, S., and Iozzo, R. V. (2004). Biologically active decorin is a monomer in solution. *J. Biol. Chem.* **279**, 6606–6612.

Hart, R. W., and Farrell, R. A. (1969). Light scattering in the cornea. *J. Opt. Soc. Am.* **59**, 766–774.

Hassell, J. R., Cintron, C., Kublin, C., and Newsome, D. A. (1983). Proteoglycan changes during restoration of transparency in corneal scars. *Arch. Biochem. Biophys.* **222**, 362–369.

Hassell, J. R., Newsome, D. A., Krachmer, J. H., and Rodrigues, M. M. (1980). Macular corneal dystrophy: Failure to synthesize a mature keratan sulfate proteoglycan. *Proc. Natl. Acad. Sci. USA* **77**, 3705–3709.

Hayashida, Y., Akama, T. O., Beecher, N., Lewis, P., Young, R. D., Meek, K. M., Kerr, B., Hughes, C. E., Caterson, B., Tanigami, A., Nakayama, J., Fukada, M. N., *et al.* (2006). Matrix morphogenesis in cornea is mediated by the modification of keratan sulfate by GlcNAc 6-*O*-sulfotransferase. *Proc. Natl. Acad. Sci. USA* **103**, 13333–13338.

Hayes, S., Boote, C., Lewis, J., Sheppard, J., Abahussin, M., Quantock, A. J., Purslow, C., Votruba, M., and Meek, K. M. (2007). Comparative study of fibrillar collagen arrangement in the corneas of primates and other mammals. *Anat. Rec.* **290**, 1542–1550.

Hodson, S., and Miller, F. (1976). The bicarbonate ion pump in the endothelium which regulates the hydration of rabbit cornea. *J. Physiol.* **263**, 563–577.

Hogan, M. J., Alvarado, J. A., and Weddell, J. E. (1971). Histology of the human eye. "An Atlas and Textbook." Chapter 3. The Cornea. pp. 55–111. Saunders, Philadelphia.

Holmes, D. F., *et al.* (2001). Corneal collagen fibril structure in three dimensions: Structural insights into fibril assembly, mechanical properties, and tissue organization. *Proc. Natl. Acad. Sci. USA* **98**, 7307–7312.

Imberty, A., Lortat-Jacob, H., and Pérez, S. (2007). Structural view of glycosaminoglycan-protein interactions. *Carbohydr. Res.* **342**(3–4), 430–439.

Kadler, K. E., *et al.* (1996). Collagen fibril formation. *Biochem. J.* **316**, 1–11.

Kinoshita, S., Adachi, W., Sotozono, C., Nishida, K., Yokoi, N., Quantock, A. J., and Okubo, K. (2001). Characteristics of the ocular surface epithelium. *Prog. Retin. Eye Res.* **20**, 639–673.

Kjellén, L., and Lindahl, U. (1991). Proteoglycans: Structures and interactions. *Annu. Rev. Biochem.* **60**, 443–475.

Knupp, C., Amin, S. Z., *et al.* (2002a). Collagen VI assemblies in age-related macular degeneration. *J. Struct. Biol.* **139**, 181–189.

Knupp, C., Chong, N. H.V., *et al.* (2002b). Analysis of the collagen VI assemblies associated with Sorsby's fundus dystrophy. *J. Struct. Biol.* **137**, 31–40.

Knupp, C., and Squire, J. M. (2001). A new twist in the collagen story — The type VI segmented supercoil. *EMBO J.* **20**, 372–376.

Knupp, C., and Squire, J. M. (2005). Molecular packing in network-forming collagens. *Adv. Protein Chem.* **70**, 375–403.

Knupp, C., *et al.* (2006). Structural correlation between collagen VI microfibrils and collagen VI banded aggregates. *J. Struct. Biol.* **154**, 312–326.

Koga, T., *et al.* (2005). Expression of a chondroitin sulfate proteoglycan, versican (PG-M), during development of rat cornea. *Curr. Eye Res.* **30**, 455–463.

Komai, Y., and Ushiki, T. (1991). The three-dimensional organization of collagen fibrils in the human cornea and sclera. *Invest. Ophthalmol. Vis. Sci.* **32**, 2244–2258.

Li, W., Vergnes, J.-P., Cornuet, P. K., and Hassell, J. R. (1992). cDNA clone to chick corneal chondroitin dermatan sulfate proteoglycan reveals identity to decorin. *Arch. Biochem. Biophys.* **296**, 190–197.

Linsenmayer, T. F. (1991). Collagens. *In* "Cell Biology of Extracellular Matrix" (E. D. Hay, ed.), 2nd edn., pp. 6–44. Plenum Press, New York.

Liu, C.-Y., Shiraishi, A., Kao, C. W., Converse, R. L., Funderburgh, J. L., Corpus, L. M., Conrad, G. W., and Kao, W. W. (1998). The cloning of mouse keratocan cDNA and genomic DNA and the characterization of its expression during eye development. *J. Biol. Chem.* **273**, 22584–22588.

Marshall, G. E., Konstas, A. G., and Lee, W. R. (1991). Immunogold fine structural localization of extracellular matrix components in aged human cornea. I. Types I–IV collagen and laminin. *Graefe's Arch. Clin. Exp. Ophthalmol.* **229**, 157–163.

Marshall, G. E., Konstas, A. G., and Lee, W. R. (1993). Collagens in ocular tissues. *Br. J. Ophthalmol.* **77**, 515–524.

Maurice, D. M. (1957). The structure and transparency of the cornea. *J. Physiol.* **136**, 263–286.

Maurice, D. M. (1962). Clinical physiology of the cornea. *Int. Ophthalmol. Clin.* **2**, 561–572.

Maurice, D. M. (1972). The location of the fluid pump in the cornea. *J. Physiol.* **221**, 43–54.

Meek, K. M., and Boote, C. (2009). The use of X-ray scattering techniques to quantify the orientation and distribution of collagen in the corneal stroma. *Prog. Retin. Eye Res.*, **28**(5), 369–392.

Meek, K. M., and Leonard, D. W. (1993). Ultrastructure of the corneal stroma: A comparative study. *Biophys. J.* **64**, 273–280.

Meek, K. M., Quantock, A. J., Elliott, G. F., Ridgway, A. E., Tullo, A. B., Bron, A. J., and Thonar, E. J. (1989). Macular corneal dystrophy: The macromolecular structure of the stroma observed using electron microscopy and synchrotron x-ray diffraction. *Exp. Eye Res.* **49**, 941–958.

Meek, K. M., *et al.* (1981). Interpretation of the meridional x-ray diffraction pattern from collagen fibrils in corneal stroma. *J. Mol Biol.* **149**, 477–488.

Meek, K. M., *et al.* (1987). The organisation of collagen fibrils in the human corneal stroma: A synchrotron X-ray diffraction study. *Curr. Eye Res.* **6**, 841–846.

Morishige, N., Wahlert, A. J., Kenney, M. C., Brown, D. J., Kawamoto, K., Chikama, T., Nishida, T., and Jester, J. V. (2007). Second-harmonic imaging microscopy of normal human and keratoconus cornea. *Invest. Ophthalmol. Vis. Sci.* **48**, 1087–1094.

Müller, L. J., *et al.* (2004). A new three-dimensional model of the organization of proteoglycans and collagen fibrils in the human corneal stroma. *Exp. Eye Res.* **78**, 493–501.

Nakamura, K. (2003). Interaction between injured corneal epithelial cells and stromal cells. *Cornea* **22**(7 Suppl.), S35–S47.

Newsome, D. A., Gross, J., and Hassell, J. R. (1982). Human corneal stroma contains three distinct collagens. *Invest. Ophthalmol. Vis. Sci.* **22**, 376–381.

Newton, R. H., and Meek, K. M. (1998). Circumcorneal annulus of collagen fibrils in the human limbus. *Invest. Ophthalmol. Vis. Sci.* **39**, 1125–1134.

Okuyama, K. (2008). Revisiting the molecular structure of collagen. *Connect. Tissue Res.* **49**, 299–310.

Poole, C. A., Ayad, S., and Gilbert, R. T. (1992). Chondrons from articular cartilage. V. Immunohistochemical evaluation of type VI collagen organisation in isolated chondrons by light, confocal and electron microscopy. *J. Cell Sci.* **103**, 1101–1110.

Prakash, G., *et al.* (2009). Comparison of fourier-domain and time-domain optical coherence tomography for assessment of corneal thickness and intersession repeatability. *Am. J. Ophthalmol.* **148**, 282–290.

Prockop, D. J., and Kivirikko, K. I. (1995). Collagens: Molecular biology, diseases, and potentials for therapy. *Annu. Rev. Biochem.* **64**, 403–434.

Quantock, A. J., Meek, K. M., Ridgway, A. E., Bron, A. J., and Thonar, E. J. (1990). Macular corneal dystrophy: Reduction in both corneal thickness and collagen interfibrillar spacing. *Curr. Eye Res.* **9**, 393–398.

Quantock, A. J., and Young, R. D. (2008). Development of the corneal stroma, and the collagen-proteoglycan associations that help define its structure and function. *Dev. Dyn.* **237**, 2607–2621.

Rada, J., Cornuet, P. K., and Hassell, J. R. (1993). Regulation of corneal collagen fibrillogenesis in vitro by corneal proteoglycan (lumican and decorin) core proteins. *Exp. Eye Res.* **56**, 635–648.

Raspanti, M., *et al.* (2008). Glycosaminoglycans show a specific periodic interaction with type I collagen fibrils. *J. Struct. Biol.* **164**, 134–139.

Reale, E., *et al.* (2001). In the mammalian eye type VI collagen tetramers form three morphologically different aggregates. *Matrix Biol.* **20**, 37–51.

Ricard-Blum, S., and Ruggiero, F. (2005). The collagen superfamily: From the extracellular matrix to the cell membrane. *Pathol. Biol.* **53**(7), 430–442.

Rich, A., and Crick, F. H. (1961). The molecular structure of collagen. *J. Mol. Biol.* **3**, 483–506.

Scott, J. E. (1992a). Morphometry of cupromeronic blue-stained proteoglycan molecules in animal corneas, versus that of purified proteoglycans stained in vitro, implies that tertiary structures contribute to corneal ultrastructure. *J. Anat.* **180**, 155–164.

Scott, J. E. (1992b). Supramolecular organization of extracellular matrix glycosaminoglycans, in vitro and in the tissues. *FASEB J.* **6**, 2639–2645.

Scott, J. E. (2001). Structure and function in extracellular matrices depend on interactions between anionic glycosaminoglycans. *Pathol. Biol.* **49**, 284–289.

Scott, J. E. (2003). Elasticity in extracellular matrix 'shape modules' of tendon, cartilage, etc. A sliding proteoglycan-filament model. *J. Physiol.* **553**, 335–343.

Scott, J. E., and Bosworth, T. R. (1990). A comparative biochemical and ultrastructural study of proteoglycan–collagen interactions in corneal stroma. Functional and metabolic implications. *Biochem. J.* **270**, 491–497.

Scott, J. E., and Haigh, M. (1985). 'Small'-proteoglycan:collagen interactions: Keratan sulphate proteoglycan associates with rabbit corneal collagen fibrils at the 'a' and 'c' bands. *Biosci. Rep.* **5**, 765–774.

Scott, J. E., and Haigh, M. (1988). Keratan sulphate and the ultrastructure of cornea and cartilage: A 'stand-in' for chondroitin sulphate in conditions of oxygen lack? *J. Anat.* **158**, 95–108.

Scott, J. E., and Parry, D. A. (1992). Control of collagen fibril diameters in tissues. *Int. J. Biol. Macromol.* **14**, 292–293.

Scott, P. G., *et al.* (2004). Crystal structure of the dimeric protein core of decorin, the archetypal small leucine-rich repeat proteoglycan. *Proc. Natl. Acad. Sci. USA* **101**, 15633–15638.

Shuttleworth, C. A. (1997). Type VIII collagen. *Int. J. Biochem. Cell Biol.* **29**, 1145–1148.

Smith, T. B. (1988). Multiple scattering in the cornea. *J. Mod. Opt.* **35**, 93–101.

van der Rest, M., and Garrone, R. (1991). Collagen family of proteins. *FASEB J.* **5**, 2814–2823.

von der Mark, H., *et al.* (1984). Immunochemistry, genuine size and tissue localization of collagen VI. *Eur. J. Biochem.* **142**, 493–502.

Wess, T. J. (2005). Collagen fibril form and function. *Adv. Protein Chem.* **70**, 341–374.

Wilson, S. E., and Hong, J. W. (2000). Bowman's layer structure and function: Critical or dispensable to corneal function? A hypothesis. *Cornea* **19**, 417–420.

Wu, J. J., and Eyre, D. R. (1995). Structural analysis of cross-linking domains in cartilage type XI collagen. Insights on polymeric assembly. *J. Biol. Chem.* **270**, 18865–18870.

Young, B. B., *et al.* (2002). The roles of types XII and XIV collagen in fibrillogenesis and matrix assembly in the developing cornea. *J. Cell. Biochem.* **87**, 208–220.

Zhang, G., Chen, S., Goldoni, S., Calder, B. W., Simpson, H. C., Owens, R. T., McQuillan, D. J., Young, M. F., Iozzo, R. V., and Birk, D. E. (2009). Genetic evidence for the coordinated regulation of collagen fibrillogenesis in the cornea by decorin and biglycan. *J. Biol. Chem.* **284**, 8888–8897.

STRUCTURAL BIOLOGY OF PERIPLASMIC CHAPERONES

By WILLIAM J. ALLEN, GILLES PHAN, AND GABRIEL WAKSMAN

Institute of Structural and Molecular Biology, Birkbeck and University College London, Malet Street, London WC1E 7HX, UK

ABSTRACT

Proteins often require specific helper proteins, chaperones, to assist with their correct folding and to protect them from denaturation and aggregation. The cell envelope of Gram-negative bacteria provides a particularly challenging environment for chaperones to function in as it lacks readily available energy sources such as adenosine $5'$ triphosphate (ATP) to power reaction cycles. Periplasmic chaperones have therefore evolved specialized mechanisms to carry out their functions without the input of external energy and in many cases to transduce energy provided by protein folding or ATP hydrolysis at the inner membrane.

Structural and biochemical studies have in recent years begun to elucidate the specific functions of many important periplasmic chaperones and how these functions are carried out. This includes not only specific carrier chaperones, such as those involved in the biosynthesis of adhesive fimbriae in pathogenic bacteria, but also more general pathways including the periplasmic transport of outer membrane proteins and the extracytoplasmic stress responses. This chapter aims to provide an overview of protein chaperones so far identified in the periplasm and how structural biology has assisted with the elucidation of their functions.

I. INTRODUCTION

The cell envelope of Gram-negative bacteria consists of two membranes separated by a narrow aqueous space called the periplasm. The inner membrane is impermeable to most solutes, with traffic of ions, proteins, and small molecules mediated by specific inner membrane proteins and highly regulated. In contrast, the outer membrane contains a large number of integral pore proteins, porins, which allow the free passage of water and small hydrophilic molecules. The outer membrane thus allows influx of many nutrients while providing an effective barrier to the transport of small hydrophobic molecules and macromolecules larger than about 600 Da (Bos *et al.*, 2007; Nikaido, 2003). Because of this selective permeability of the outer membrane, many aspects of environment in the periplasm closely resemble those of the external medium: it is generally oxidizing, has salt and pH levels similar to those outside the cell, and does not contain the usual energy sources available in the cytosol, such as adenosine triphosphate (ATP). In addition, a layer of peptidoglycan within the periplasm, along with a high concentration of protein, probably makes the periplasmic space highly crowded and viscous (Mogensen and Otzen, 2005; Wulfing and Pluckthun, 1994).

Many crucial bacterial functions take place within the cell envelope, relying on the ability of bacteria to import molecules such as nutrients from the external medium, to export effector molecules synthesized in the cytoplasm, or to pump out toxic molecules such as antibiotics. For example, pathogenic bacteria often use adhesin proteins anchored in their outer membranes to mediate attachment to their target cells, and they are able to secrete toxins into host cells to facilitate colonization and infection. Since all bacterial proteins are synthesized in the cytoplasm, their correct localization to the periplasm, outer membrane, or external medium requires specialized export machineries. Export of numerous effector molecules is carried out by specialized bacterial secretion systems (Durand *et al.*, 2009); insertion of integral outer membrane proteins into the outer membrane is assisted by the conserved β-barrel assembly machinery (BAM) (Knowles *et al.*, 2009); and association of membrane-anchored lipoproteins with the outer membrane requires the lipoprotein localization (Lol) complex (Tokuda, 2009). Integral to all these processes are a number of molecular chaperones — proteins that assist in the folding, assembly, and stabilization of other proteins or complexes, preventing undesirable protein–protein interactions and guiding their targets to their correct conformation and location.

Chaperones in the periplasm can be divided broadly into two basic functions with a certain amount of functional overlap. Carrier chaperones, as typified by PapD-like chaperones in the chaperone/usher pathway of pilus biogenesis, are involved in the stabilization of specific substrates within the periplasm prior to assembly or export. This group of chaperones sequester their substrates away from the general environment, protecting them from aggregation or degradation and often targeting them to their final location within the cell. Because of the lack of external energy sources in the periplasm, these chaperones are often able to store energy from substrate binding and utilize it later to power transport or reactivity. Chaperones that hold their substrates in metastable configurations competent for activity in this manner have also been termed "steric chaperones," as they provide steric information to their substrates rather than acting as passive holding devices (Pauwels *et al.*, 2007). The second group of periplasmic chaperones, members of which assist with the correct folding of other periplasmic proteins and protect them from aggregation and proteolysis, includes the chaperones SurA, Skp, and DegP. These proteins are generally upregulated in response to one or both of the envelope stress pathways σ^E and Cpx, and are able to bind a variety of different unfolded or misfolded proteins (Duguay and Silhavy, 2004; Sklar *et al.*, 2007). This chapter discusses periplasmic chaperones so far identified as being involved in protein folding, stabilization, or transport. In particular, we explore how the structures of individual proteins allow them to carry out their specific functions in the absence of ATP or other convenient energy sources.

II. Chaperones Involved in Folding and Stabilization of Proteins in the Periplasm

Most chaperones discovered to date were initially identified as proteins upregulated in response to elevated temperature. Heat shock, along with other stress factors such as changes in pH or the presence of ethanol, destabilizes the native forms of proteins, making them prone to unfolding, misfolding, and aggregation (Duguay and Silhavy, 2004). Since aggregation is a highly favorable process within the crowded environment of the cytoplasm or periplasm, and buildup of aggregated protein leads to cell damage or death, cells have evolved specific mechanisms to combat this effect. Stress response pathways, which are activated in response to a variety of stress conditions, stimulate production or activation of factors that help fold and protect proteins along with proteases to degrade irreversibly damaged proteins. Two main pathways activate in response to periplasmic and membrane stress in *Escherichia coli*: σ^E and Cpx. These pathways are thought to be stimulated by the presence of unfolded protein in the periplasm and upregulate a different but overlapping set of genes encoding periplasm-specific folding factors and proteases (Duguay and Silhavy, 2004; Rhodius *et al.*, 2006).

Although most of the chaperones responsible for general folding and stabilization in the periplasm are upregulated in response to cell envelope stress, many of them have essential roles even under normal conditions. The majority of proteins destined for the periplasm, the outer membrane, or secretion are translocated across the inner membrane by the general secretory (Sec) pathway and reach the periplasm as unfolded polypeptide chains (Driessen and Nouwen, 2008). Because spontaneous protein folding can be a slow and error prone process, these newly synthesized proteins often make use of chaperones and other folding factors within the periplasm to reach their final folded states. Integral outer membrane proteins in particular require assistance for correct assembly as their native β-barrel structures expose large hydrophobic areas on the surface. This is necessary for their stability once embedded in the lipid environment of the membrane; however, it makes them highly unstable and susceptible to aggregation after exit from the Sec translocon and during transit through the aqueous periplasm (Bos *et al.*, 2007; Knowles *et al.*, 2009).

Three particular periplasmic chaperones in *E. coli* — SurA, Skp, and DegP — have been implicated as being specifically responsible for carrying the outer membrane proteins through the periplasm to the outer membrane (Sklar *et al.*, 2007). Spontaneous membrane insertion of β-barrels is a very slow process, as it requires several hydrophilic loops between the β-strands to pass through the hydrophobic membrane to the extracellular side. For most outer membrane proteins, this step thus requires assistance from the conserved BAM, which is thought to orchestrate both folding and membrane insertion, a complex multistep process (for a recent review see Knowles *et al.*, 2009). Chaperones involved in this process would therefore need to possess dual functionality: the ability to shuttle their substrates across the periplasm in a state competent for insertion and the ability to target their substrates to the BAM complex and hand them over for the insertion step.

A. *Protein Folding in the Periplasm*

Cells use three different types of folding catalyst to assist with general protein folding: chaperones, which bind to proteins in stoichiometric quantities and assist folding by preventing off-pathway interactions; peptidyl–prolyl isomerases (PPIases), which catalyze the *cis-trans* isomerization of proline residues in proteins' structures, thought to be a limiting step in the rate of folding of many proteins (Justice *et al.*, 2005); and protein disulfide isomerases, which catalyze the formation and exchange of disulfide bonds (Inaba, 2009). This latter category is largely specific to cell envelope and secreted proteins, as disulfide bonds are not stable in the reducing environment of the cytoplasm. The majority of periplasmic chaperones so far identified as being involved in general folding of proteins have both chaperone function and other stress response or protein folding roles, encoded on the same polypeptide chain but often at different sites or on different domains. For example, SurA and FkpA, two of the best-characterized periplasmic chaperones, also have PPIase activity and in both cases this catalytic activity can be specifically knocked out without affecting the chaperone function (Behrens *et al.*, 2001; Ramm and Pluckthun, 2000). DegP, another important component of both the σ^E and Cpx stress responses, has dual protease and chaperone functions and can switch between the two depending on whether its substrate is correctly folded or misfolded (Krojer *et al.*, 2008). And some of the periplasmic disulfide bond isomerases, including DsbA, DsbC, and DsbG in *E. coli*, have also been shown to prevent aggregation of unfolded proteins *in vitro* (Chen *et al.*, 1999; Shao *et al.*, 2000; Zheng *et al.*, 1997).

In addition to the functional duality, many periplasmic chaperones have partial functional redundancy and are able to compensate if other chaperones are missing. As a result, knocking out individual chaperones often results in very minor effects on the cell growth. For example, SurA is one of the most important chaperones for assisting the periplasmic transit of outer membrane proteins; however, knocking this protein out is not lethal to the cell, leading only to minor outer membrane protein defects, characterized by susceptibility to hydrophobic molecules such as detergents (Justice *et al.*, 2005; Rizzitello *et al.*, 2001). This ability to tolerate the deletion of a major folding factor arises because other proteins, in this case Skp and DegP, have partially overlapping function and are able to pick up some of the slack. Indeed, double knockouts of SurA/ Skp or SurA/DegP are synthetically lethal, providing direct evidence that they work on parallel pathways (Rizzitello *et al.*, 2001). Interestingly, the Skp/DegP double mutation is not lethal, which was taken as evidence that these two proteins function within the same pathway for outer membrane transport. This result was subsequently replicated using depletion rather than deletion studies, and it was concluded that SurA carries out the bulk of periplasmic shuttling under normal circumstances, with Skp/DegP primarily dealing with proteins that fall off the normal folding pathway (Sklar *et al.*, 2007).

Efforts to disentangle the roles of individual chaperones from the complex network of other folding catalysts are further hampered by the fact that many knockouts activate the periplasmic stress responses, which then upregulate the synthesis of other chaperones and downregulate the production of many outer membrane proteins. The phenotypes observed in genetic experiments involving the periplasmic chaperones are therefore often highly pleiotropic (Duguay and Silhavy, 2004; Mogensen and Otzen, 2005). Because of this difficulty in assigning specific roles to specific proteins, there are still several periplasmic folding factors whose chaperone functions are poorly understood. It is also quite possible that additional, as of yet undiscovered chaperones exist in the periplasm and contribute to protein folding in unknown ways. This is especially true of bacterial species that are less well characterized than *E. coli*, particularly those that reside in different environments and are thus subject to different stresses, requiring different stress response factors (Rhodius *et al.*, 2006). Despite these complications, some of the periplasmic chaperones — notably those involved in the transport of integral outer membrane proteins — have been fairly well studied. Genetic studies, biochemistry, structural biology, and biophysical methods have together yielded a decent understanding of the roles in particular of Skp, SurA, and DegP and how they accomplish their tasks.

B. SurA

SurA is a periplasmic chaperone initially identified as a survival factor in stationary phase cells (Tormo *et al.*, 1990), although this phenotype only manifests itself under certain conditions. The primary function of SurA *in vivo* is thought to be the folding and periplasmic shuttling of integral outer membrane proteins. A number of results support this notion: first, the phenotypes of SurA deletion are characteristic of outer membrane defects, in particular the sensitivity to hydrophobic compounds which suggests increased permeability of the outer membrane (Justice *et al.*, 2005). Deletion of SurA also directly affects the amount of protein present in the outer membrane, both in terms of total membrane density (Sklar *et al.*, 2007) and specific outer membrane proteins (Rouviere and Gross, 1996). Screening of peptide libraries identified a high affinity of SurA for sequences rich in aromatic residues, in particular contained within the motif of Ar-X-Ar, where Ar is an aromatic residue and X is polar. This pattern is a common feature of integral outer membrane proteins, as amphipathic beta strands are characterized by patterns of alternating hydrophilic and hydrophobic residues. Ar-X-Ar is particularly prevalent at the periplasmic turns of such proteins, as aromatic residues tend to cluster at the ends of the β-strands that make up β-barrels (Bitto and McKay, 2003; Hennecke *et al.*, 2005). In addition, SurA binds with much higher affinity to unfolded outer membrane proteins than to unfolded soluble proteins or any folded protein (Bitto and McKay, 2004).

The crystal structure of the entire SurA polypeptide comprises four domains: an N-terminal domain with no known structural homologues (blue in Fig. 1A), two central domains with the conserved parvulin fold observed in several PPIases (P1 and P2; green and yellow, respectively, in Fig. 1A), and a short C-terminal domain that forms a single, long α-helix followed by a short β-strand (red in Fig. 1A; Bitto and McKay, 2002). The P2 domain, which contains the PPIase activity (Rouviere and Gross, 1996), forms a satellite domain about 30 Å removed from the other three domains, which together form the core of the protein.

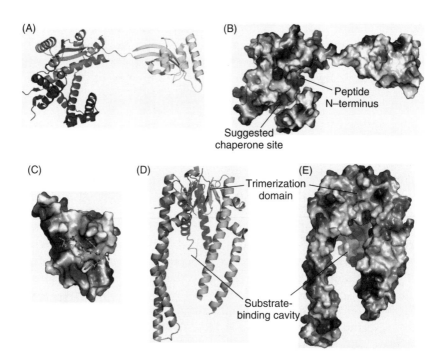

Fɪɢ. 1. Structures of outer membrane protein chaperones SurA and Skp (PDB code 1SG2). (A) Cartoon representation of full-length SurA [PDB code 1M5Y (Bitto and McKay, 2002)], with the N-terminal domain colored blue, the P1 domain green, the P2 domain yellow, and the C-terminal domain red. (B) Surface model of full-length SurA in the same orientation as in (A), colored by a gradient of electrostatic potential from negative (red) to positive (blue). A substrate peptide (WEYIPNV) bound to the P1 pocket is shown in stick representation with carbons in green, nitrogens in blue, and oxygens in red. The orientation of this peptide was determined by superposition of the crystal structure of the P1 domain with bound peptide [PDB code 2PV1 (Xu *et al.*, 2007)] onto the P1 domain of the full-length structure. (C) Surface model of the P1 domain of SurA in complex with the model peptide, using the same display mode as in (B). (D) Cartoon structure of Skp [PDB code 1SG2 (Walton and Sousa, 2004)], with the three subunits forming the trimer colored in cyan, magenta, and orange, respectively. (E) Surface model of Skp in the same orientation as in (D), colored by a gradient of electrostatic potential from negative (red) to positive (blue). (See Color Insert.)

This structural separation between the core and satellite domains mirrors a functional division between the PPIase activity, carried out entirely by the P2 domain (Rouviere and Gross, 1996), and the chaperone activity, which is barely affected by deletion of P2 (Behrens *et al.*, 2001).

A deep groove in the surface of the N-terminal domain of SurA was originally proposed to be the site of substrate binding (Fig. 1B). In the original crystal structure, this site is occupied by peptides from adjacent SurA molecules in the crystal packing arrangement, showing a potential mode of binding (Bitto and McKay, 2002). Further support for the idea that the main substrate-binding site is on the N-terminal domain came from the observation that the observable SurA deletion phenotype can be suppressed by a truncated form of SurA comprising only the N- and C-terminal domains (Behrens *et al.*, 2001; Watts and Hunstad, 2008) and that the N-terminal domain alone is able to bind model peptides (Webb *et al.*, 2001). This suggests that the P1 domain has only a marginal role in SurA activity, and indeed some homologues of SurA from other species lack the P1 domain entirely (Bitto and McKay, 2002). However, subsequent crystal structures of SurA in complex with consensus substrate peptides show them binding to the P1 domain, with specific pockets present for the aromatic residues (Fig. 1C; Xu *et al.*, 2007). As pointed out by Xu *et al.*, to reconcile these conflicting results it seems necessary to invoke separate substrate recognition and chaperone activity sites, the former being on the P1 domain and responsible for selecting outer membrane proteins over general unfolded proteins and the latter able to protect these substrates from the aqueous environment of the periplasm and target them to the outer membrane. Examination of the positions of the two sites reveals that the N-terminus of the peptide bound to the "recognition" site of P1 is close to one end of the proposed "chaperoning" site on the N-terminal domain (Fig. 1B). This raises the intriguing possibility that a longer peptide, such as in a full-length protein, could bind to both sites at once, perhaps with the Ar-X-Ar motif occupying the recognition site and the subsequent β-strand of the barrel occupying the chaperone site. Indeed, it has been noted that the distance between the SurA consensus binding sites on several important outer membrane proteins is roughly 10–20 residues, just right to fit into the 50-Å binding channel provided by the crevice in the N-terminal domain (Xu *et al.*, 2007).

Despite the well-characterized substrate interactions of SurA, it is still not clear exactly which stages of the BAM process the chaperone mediates. Functional studies on the homotrimeric pore protein LamB, in which the individual subunits are folded separately prior to trimerization and insertion into the membrane suggest that the role of the protein is primarily folding of the monomers and that trimerization and membrane insertion are catalyzed at a later stage by the outer membrane BAM complex (Rouviere and Gross, 1996). However, SurA is found to fractionate primarily with the outer membrane (Hennecke *et al.*, 2005), and the effect of its deletion on LamB synthesis is indistinguishable from the effect of deleting the *E. coli* protein YgfL, a component of the outer membrane BAM complex. This would give SurA a role in transporting its substrates to the outer membrane and localizing them to the BAM complex directly, but not in the actual folding itself (Ureta *et al.*, 2007). This latter suggestion is consistent with the structural model described above, where the β-strands in the unfolded protein are kept separate by SurA, which must dissociate to allow them to come together into the final β-barrel tertiary or quaternary structure. Thus, it seems likely that the role of SurA is to bind to unfolded outer membrane proteins as or immediately after they dissociate from the Sec translocon and transport them in a stable unfolded state to the outer membrane, where they are assembled into their native structure by the BAM complex.

C. Skp

Skp (17 kDa protein) is a second periplasmic chaperone involved in the stabilization of outer membrane proteins during passage through the periplasm (Chen and Henning, 1996). Consistent with this, it has been shown to interact specifically with unfolded outer membrane proteins as or immediately after they are released from the inner membrane Sec complex and assist with their release from the inner membrane (Harms *et al.*, 2001; Schafer *et al.*, 1999). The binding of Skp allows the hydrophobic, aggregation-prone β-barrel proteins to remain stable while traversing the periplasm and is thought to assist with their insertion into the outer membrane — or at least tranfer them to the Bam complex, which would then carry out the membrane insertion step. Indeed, it has been shown that Skp and lipopolysaccharide (LPS), a major component of the outer leaflet of the outer membrane, are together sufficient to mediate insertion of the β-barrel protein into lipid bilayers (Bulieris *et al.*, 2003). Skp forms a stable trimer in solution, and this trimer binds its substrate in a strict 1:1 ratio (one trimer of Skp per molecule of substrate protein).

The crystal structure of the Skp trimer (Korndorfer *et al.*, 2004; Walton and Sousa, 2004) has been likened to a jellyfish, with a central β-barrel "body" trimerization domain — to which each subunit contributes four β-strands — and long straggly α-helical tentacles reaching out to grasp substrates (Fig. 1D). The three subunits superimpose well at the body, which forms a rigid hydrophobic core; however, the tentacles are highly flexible, with hinge loops at the point where they meet the body, allowing substantial movement. The α-helical arms have a net positive charge on their outside surface — thought to provide a binding site for LPS — and hydrophobic patches on their inside faces, presumably responsible for binding hydrophobic substrates (Fig. 1E). Together the arms form a central cavity large enough to hold an outer membrane pore, with different-sized proteins presumably accommodated by the flexibility of the hinges (Walton and Sousa, 2004).

Recent fluorescence (Qu et al., 2009) and nuclear magnetic resonance (NMR) (Walton et al., 2009) studies have clarified the interaction of Skp with one of its major substrates, the monomeric integral outer membrane protein OmpA, which possesses both a membrane-inserted β-barrel domain and a soluble periplasmic domain. By analyzing the environment and mobility of various residues at different points within the substrate structure, the above studies were able to show that the β-barrel domain of OmpA collapses to an unfolded but inside-out structure within the central cavity of Skp, with contacts not only between hydrophobic patches on the two proteins but importantly also between positively charged regions on the Skp tentacles and negatively charged patches in the loop regions of OmpA (Qu et al., 2007). The soluble periplasmic domain of OmpA, meanwhile, protrudes into the solvent and folds independently. Binding of LPS weakens the Skp:OmpA interaction and causes some of the surface loops to become more solvent exposed, likely as the negatively charged LPS competes with OmpA for the binding sites on Skp. This is necessary but not sufficient for efficient folding of OmpA by Skp *in vitro* — lipid bilayers are ultimately required for concomitant folding and insertion (Bulieris et al., 2003). It should be noted, however, that the nature of the final insertion step under cellular conditions is not clear and that the BAM complex is thought to be involved in the insertion of most outer membrane proteins. It could be, therefore, that the LPS binding is not a necessary step in the *in vivo* insertion pathway mediated by Skp or that it only becomes involved during the handover to BamA at the outer membrane (Mogensen and Otzen, 2005).

OmpA in its interaction with Skp is likely to be representative of a variety of different integral outer membrane β-barrels, as the hydrophobic regions and negatively charged loops are common features of this class of protein. It is of particular interest to compare the proposed functional models of SurA and Skp, as they accomplish a similar task using very different methods. SurA binds at many sites along the length of unfolded membrane proteins, recognizing specific primary or secondary structural motifs. Conversely, the Skp trimer encloses entire molten globules of unfolded membrane proteins in a 1:1 stoichiometry, recognizing a tertiary pattern of hydrophobic and negatively charged regions (Qu *et al.*, 2007). In both cases, however, the chaperones keep their substrates unfolded and transport them to the outer membrane for insertion. If, as has been observed for spontaneous *in vitro* β-barrel formation (Kleinschmidt *et al.*, 1999; Qu *et al.*, 2009), membrane insertion is inextricably linked to folding — perhaps with the energy released by the folding process required to power the insertion step — then it seems that the ability to keep membrane proteins soluble while preventing them from folding prematurely would be a key function of periplasmic chaperones.

D. DegP

DegP is the *E. coli* periplasmic representative of the high temperature requirement (HtrA) family of proteins, which promote survival at elevated temperatures. This survival is mediated by two separate functions: a chaperone activity that rescues slightly misfolded proteins and a protease activity that rapidly degrades irretrievably misfolded proteins (Clausen *et al.*, 2002). At low temperatures (28 °C), the chaperone function is dominant and the protease activity substantially attenuated, but as the temperature increases, substrates encountering DegP are less likely to fold and more likely to be degraded (Spiess *et al.*, 1999). This is probably due to an increase in the protease functionality at high temperatures, which alters partitioning between the two activities, rather than due to a decrease in the chaperone activity at high temperatures (Skorko-Glonek *et al.*, 2007). DegP is thought to function on a wide variety of substrates and is critical for survival above 37 °C or under conditions of cellular stress, such as when the chaperone SurA is knocked out or when proteins destined for the outer membrane are overexpressed (Mogensen and Otzen, 2005; Spiess *et al.*, 1999).

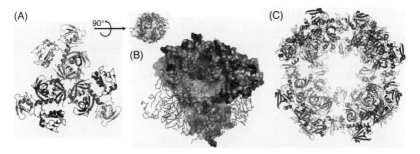

Fig. 2. Structures of DegP in various assembly states. (A) Cartoon representation of a DegP trimer from the original hexamer structure [PDB code 1KY9 (Krojer *et al.*, 2002)], viewed from above with the protease domains colored red, the PDZ1 domains in blue, and the PDZ2 domains in green. (B) Side view of the DegP hexamer, with three of the subunits (two from the trimer at the top and one from the trimer at the bottom) shown colored by a gradient of electrostatic potential from negative (red) to positive (blue). The remaining three subunits are shown in backbone ribbon representation, using the same color scheme as in (A). (C) Twelve of the DegP subunits from the 24-mer crystal structure [PDB code 3CS0 (Krojer *et al.*, 2008)], looking into the inside of the cavity and shown using the same display mode as (A). (See Color Insert.)

The DegP monomer is made up of three domains: at the N-terminus a protease domain with homology to the trypsin family of serine proteases (red in Fig. 2A) and at the C-terminal end two consecutive PSD95, DlgA, ZO-1 (PDZ) domains (blue and green, respectively, in Fig. 2A) — PDZ being a common protein module involved in protein–protein interactions. The minimal active unit of DegP is a trimer (DegP$_3$), with tight hydrophobic interactions between the protease domains forming the interface and the PDZ domains sticking out from the sides with a highly flexible orientation (see Fig. 2A; Clausen *et al.*, 2002; Krojer *et al.*, 2002). This trimer is the smallest unit normally observed for DegP and displays both high protease activity and normal chaperone activity. However, the trimer is only thought to be a major form at elevated temperatures; at lower temperatures these trimers associate to form larger complexes, with the extent of multimerization dependent on the presence and nature of substrate (Krojer *et al.*, 2008).

In the absence of substrate, the most common form of DegP is a hexamer (DegP$_6$), formed by the association of two trimers (Fig. 2B). As revealed by the first crystal structure of DegP, which captured the hexameric form (Krojer *et al.*, 2002), the association between the two trimers is mediated by the PDZ domains. These interacting domains form the walls of a large cage, with three protease modules located at either end with their active sites facing inward. The conformation of the PDZ domains within the hexamers is flexible, and they are able to open up to admit small or unfolded protein substrates into the central cavity or close down to seal off access. Not only are the protease active sites sequestered inside the cage in the hexameric form, and thus inaccessible except via entry past the PDZ domains, but they are also partially occluded by loops from adjacent subunits (Krojer *et al.*, 2002). It therefore seems plausible that the main purpose of the oligomerization of $DegP_3$ units might be the regulation of substrate access to the protease domains. *In vitro* assays comparing the activity of wild-type DegP and a mutant form unable to associate as hexamers reveal that the chaperone activity is unaltered by the mutation, whereas the protease activity is reduced (Jomaa *et al.*, 2007). Thus, the hexameric form would be less prone to indiscriminately proteolyzing undamaged proteins and therefore more economical in situations where aggregated protein levels are not critical.

While the hexamer is the most stable form of DegP in the absence of any substrate, further oligomerization into 12-mers or 24-mers can be observed upon addition of denatured protein. These increasing complex structures use the same base unit — a DegP trimer formed by association of the protease domains — and form similar but larger hollow shells with the protease active sites located on the inside (Fig. 2C). The main differences between the various multimeric forms are the orientation and packing of the PDZ domains (Jiang *et al.*, 2008; Krojer *et al.*, 2008). Interestingly, the inhibition of protease activity observed for the hexamer is relieved in the higher order structures, through a rearrangement of the regulatory loop. This suggested a mechanism for the regulation of DegP protease activity, whereby the presence of unfolded protein within the cavity induces the association of multiple hexamers, thereby derepressing proteolysis. Consistent with this, addition of denatured substrates such as lysozyme, casein, or albumin to purified $DegP_6$ causes a transient multimerization of the chaperone into the $DegP_{12}$ and $DegP_{24}$ forms, followed by a slow return to the hexameric state concurrent with substrate hydrolysis (Krojer *et al.*, 2008). The functional differences between the 12-mer and the 24-mer appear to be minor, and it is thought that they mainly reflect the size of the substrate within the cavity (Krojer *et al.*, 2008).

Although the chaperone activity of DegP is well documented, and its substrates include both soluble and membrane-inserted proteins, there are still some inconsistencies as to its involvement with the native folding pathway of outer membrane proteins. For example, deletion of the outer membrane-targeting phenylalanine at the C-terminus of the trimeric pore protein OmpF causes a buildup of unfolded and aggregated OmpF, which is lethal to DegP⁻ cells. Overexpression of a protease-deficient form of DegP (DegP^{S210A}) was able to alleviate this effect, suggesting that the chaperone function is at least a component of the DegP response; however, the unfolded OmpF was not incorporated correctly into the outer membrane, rather it formed monomeric, detergent-labile structures associated with both the outer and inner membranes (Misra et al., 2000). Similar results were also observed for a mutant form of OmpC unable to insert into the outer membrane due to an introduced disulfide bridge (CastilloKeller and Misra, 2003).

The above results suggest that while the DegP chaperone activity is important for the protection of cells from damage caused by misfolded outer membrane proteins, it is not able to help with their correct insertion into the outer membrane. This contrasts with the results of Krojer et al., who were able to coexpress DegP with various integral outer membrane proteins and showed the substrates to be folded within the DegP cavity, where their final tertiary structure protects them from degradation even by the active DegP protease domain. Using cryo-electron microscopy, they were also able to observe a cylindrical density within a DegP$_{12}$ cage, suggestive of a folded or nearly folded β-barrel protein (Krojer et al., 2008). This is in keeping with the results of some genetic studies, where a knockout of DegP is found to reduce outer membrane expression independently of the σ^{E} response (Krojer et al., 2008), and the observation that a DegP/SurA double knockout is synthetically lethal, but rescuable by DegP with the protease activity knocked out (Rizzitello et al., 2001). As of yet, there is no apparent mechanism through which a folded protein inside the DegP particle could be transferred to the BAM complex or the outer membrane directly for insertion. However, it is conceivable that the structures obtained by electron microscopy do not represent a fully folded outer membrane pore, but rather a collapsed semicompact folding intermediate akin to that postulated to exist within the Skp trimer. Within this context, it is provocative to note that alternate bowl-like multimerization states have recently been observed for DegP in its interaction with negatively charged lipid bilayers (Shen *et al.*, 2009), which could potentially represent a route for transfer to the outer membrane.

It should be pointed out that the proposed chaperone role for DegP — i.e., the formation of a hydrophobic cage in which membrane proteins can fold to their final conformation — is at odds with the models proposed for Skp and SurA, which deliberately do not fold their substrates so as to preserve the folding energy for membrane insertion. Since the genetic knockout and depletion studies largely point to a survival role of DegP during folding stress rather than an explicit role in outer membrane protein insertion (Rizzitello *et al.*, 2001; Sklar *et al.*, 2007), and other evidence for a direct role under normal conditions is fairly scant, it is possible that interaction of outer membrane β-barrels with DegP is by nature nonproductive and occurs only under conditions of stress. In such case, rather than indicating distinct pathways of outer membrane chaperone activity, the observed patterns of lethality for multiple knockouts or depletions could simply reflect that a certain minimum "critical mass" of combined outer membrane folding and degradation capacity is required to keep the cell functioning. As long as SurA is present at wild-type levels and there are no exacerbating conditions such as high temperature, the absence of Skp or DegP is damaging but not lethal. However, if both SurA and Skp are missing, then not enough protein is folded correctly into the outer membrane to yield viable cells and if both SurA and DegP are absent, unfolded outer membrane protein accumulates to the point where it is lethal to the cell.

E. *FkpA and Other Proline Isomerases*

Aside from SurA, discussed above, three other proteins with proline *cis–trans* isomerase functionality have been identified localizing to the periplasm of *E. coli*: FkpA, PpiA, and PpiD. These three proteins represent the three different types of PPIase fold so far identified. FkpA is a FK506-binding protein (FKBP) family protein, characterized by sensitivity to the immunosuppressant FK506, and shown to possess both PPIase and chaperone functions (Arie *et al.*, 2001). PpiA (RotA) is homologous to the cyclophilin family of PPIases, although its deletion has no discernable phenotype, and no chaperone function has hitherto been reported (Duguay and Silhavy, 2004; Kleerebezem *et al.*, 1995; Mogensen and Otzen, 2005). PpiD, like SurA, is a parvulin, first identified as a protein able to partially protect against the antibiotic sensitivity phenotype of a SurA deletion mutant (Dartigalongue and Raina, 1998). PpiD is anchored in the inner membrane by a single transmembrane helix and is thought to have some chaperone activity, binding to proteins as they emerge from the Sec translocon and by an unknown mechanism assisting their release into the periplasm (Antonoaea *et al.*, 2008). In addition to the *E. coli* proteins mentioned above, it is likely that other bacterial species encode a different set of PPIase/chaperones. For example, Par27, a parvulin PPIase and chaperone in *Bordetella pertussis*, has recently been identified from its ability to bind filamentous hemagglutinin and is thought to have a general role in chaperoning periplasmic proteins (Hodak *et al.*, 2008). This example serves as a reminder that, while *E. coli* is by far the best-characterized bacterial species, it is not necessarily representative of other types of Gram-negative bacteria.

From what is known of all these more general PPIase/chaperones, they do not appear to have a specific affinity for outer membrane proteins such as was noted for SurA. The sensitivity to hydrophobic antibiotics, characteristic of defective outer membrane transport, is observed only for SurA deletion strains and not for any other PPIase deletions, although the SurA⁻ phenotype is less severe if the other PPIases are still functional (Justice *et al.*, 2005). In addition, the substrate specificity of PpiD has been investigated and appears to be much broader than that of SurA (Stymest and Klappa, 2008), while FkpA has been shown to assist with the folding of several soluble proteins, including recombinantly expressed single-chain antibody fragments (Arie *et al.*, 2001; Bothmann and Pluckthun, 2000; Hu *et al.*, 2006). This suggests that such proteins may have a more general role in folding all types of protein in the periplasm, whether soluble or membrane associated. In fact, the general chaperone functions appear to be more important than the PPIase activity, which can often be knocked out with no particular effect (Behrens *et al.*, 2001; Kleinschmidt *et al.*, 1999). Despite the fact that PPIases are conserved in many species, *E. coli* cells are still able to grow under laboratory conditions even after all four known periplasmic PPIases have been knocked out (Justice *et al.*, 2005) — although this observation contrasts with a previous result suggesting that the SurA/PpiD double mutant is lethal (Dartigalongue and Raina, 1998). Perhaps the catalysis of proline isomerization is necessary only under conditions of cellular stress and is therefore selected for by evolution but is not always apparent in favorable laboratory conditions.

Of the general chaperones described above, only FkpA has been structurally characterized. It forms a stable homodimer, with each monomer consisting of two distinct domains with a high degree of flexibility in their relative orientations. The N-terminal domains of the two subunits are each made up of three long α-helices which entwine to form the dimer interface, while the C-terminal domains — the regions homologous to FKBP — are situated at either end of the molecule (Fig. 3A). As with SurA, the PPIase and chaperone functions are separated spatially and appear to function independently of one another, a construct comprising only the C-terminal domain folds into stable

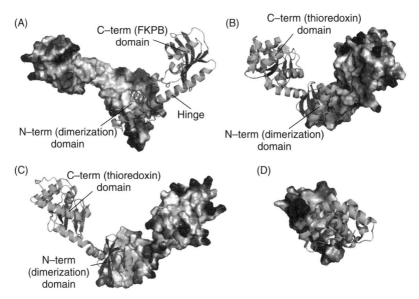

FIG. 3. Structures of various soluble chaperones involved in general purpose protein folding in the periplasm. In each case, the physiological dimers are shown with one subunit in surface representation, colored by a gradient of electrostatic potential from negative (red) to positive (blue), and the other subunit in cartoon representation, colored by secondary structure with α-helices in cyan, β-sheets in magenta, and turn regions in pink. (A) FkpA [PDB code 1Q6U (Saul *et al.*, 2004)]. In the chosen orientation, one of the three α-helices in the dimerization domain of the cartoon structure (helix 2) is obscured by the end of helix 1 in the surface structure. This is because the two domains loop through each other, giving rise to the robust dimer interface. (B) DsbC [PDB code 1EEJ (McCarthy *et al.*, 2000)]. (C) DsbG [PDB code 1V57 (Heras *et al.*, 2004)]. (D) HdeA [PDB code 1DJ8 (Gajiwala and Burley, 2000)] (See Color Insert.).

monomers and is able to catalyze PPIase activity effectively, while a truncated form of FkpA with the entire C-terminal domain removed dimerizes in solution and has been reported to display similar chaperone properties to the wild-type protein (Saul *et al.*, 2004). However, it should be noted that this latter result conflicts with earlier reports that the chaperone activity is encoded by the C-terminal domain (Ramm and Pluckthun, 2001), and it could be that both domains are involved in the chaperone function of wild-type protein.

Initially based on comparison of different crystal forms of FkpA (Saul *et al.*, 2004), the long helix connecting the two domains was proposed to be a major point of plasticity in the structure, allowing flexibility in the relative orientation of the domains (Fig. 3A). Subsequent NMR data have supported this hypothesis, and it has been proposed that this connecting arm functions as a hinge, allowing the C-terminal domains to move relatively freely with respect to each other and to the dimerization domain in the middle (Hu *et al.*, 2006). Residues on the inside surfaces of the two C-terminal domains, predominantly hydrophobic in nature and exhibiting large NMR chemical shift changes upon binding of model substrates, were predicted to form the substrate interaction site, with the distance and orientation between the two domains changing to accommodate different substrates (Hu *et al.*, 2006). It is not exactly clear from this how specific unfolded substrates are recognized, although there are suggestions that FkpA binds preferentially to oligomerized proteins, thereby preventing further aggregation (Hu *et al.*, 2006). Indeed, there does not appear to be a "specificity pocket" anywhere on the interaction surface that might select for specific amino acid sequences or three-dimensional motifs. This supports the notion that FkpA is a more general chaperone, able to bind to many different substrates provided they expose hydrophobic patches to the environment, but making only transient, nonspecific interactions with any of them.

F.   The Disulfide Bond-Forming Enzymes

An additional category of folding catalysts reported to show the chaperone activity are the disulfide bond (Dsb) isomerases, which catalyze the formation and exchange of covalent bonds between cysteine thiol groups. In contrast to proline *cis–trans* isomerization, this function is demonstrably crucial to the correct folding of many proteins and is largely periplasm-specific in bacteria, as the cytosol is maintained in a highly reducing state except under extreme stress conditions. Three soluble periplasmic Dsb proteins have been identified in the periplasm of *E. coli*, all of which have been reported to exhibit the chaperone activity in addition to their primary catalytic function [although this has been disputed for DsbA (Collet and Bardwell, 2002)]. DsbA is the main enzyme responsible for disulfide bond formation, catalyzing the oxidation of cysteines very effectively and often indiscriminately; DsbC catalyzes the exchange of disulfide bonds formed by DsbA, breaking incorrect bonds and forming bonds conducive to folding, while the physiological role of DsbG is unclear, although its overexpression can compensate for deletion of DsbC, suggesting that it performs a similar role perhaps on a different subset of substrates. The other two Dsb proteins — DsbB and DsbD — are integral inner membrane proteins that catalyze the oxidation of DsbA and reduction of DsbC/DsbG, respectively, maintaining them in their catalytic activity redox states [for recent reviews, see Inaba, 2009; Messens and Collet, 2006; Nakamoto and Bardwell, 2004].

Chaperone activities for DsbC and DsbG have been demonstrated *in vitro* using classical chaperone function assays, which monitor their ability to prevent aggregation and assist with refolding of denatured model substrates (Chen *et al.*, 1999; Shao *et al.*, 2000). This chaperoning ability is thought to be important primarily for the function of DsbC and DsbG as disulfide bond isomerases. Proteins containing incorrect disulfide bonds are likely to become misfolded, in many cases burying the mispaired cysteines within an aberrant tertiary fold. The ability to break and reform bonds must therefore be intimately connected to the rescue of denatured protein, preferably alongside the ability to specifically recognize nonnative structures (Nakamoto and Bardwell, 2004). Despite this connection between their two functions, DsbC and DsbG do not discriminate between misfolded proteins containing oxidized cysteines and other misfolded proteins that lack them (Chen *et al.*, 1999; Shao *et al.*, 2000). They are thus capable of providing general protection against folding stress within the periplasm.

The structures of DsbC (McCarthy *et al.*, 2000) and DsbG (Heras *et al.*, 2004), which have high sequence and structural homology, both bear some overall architectural resemblance to the FkpA structure described above. All three proteins form V-like shapes with N-terminal dimerization domains required for the chaperone activity and C-terminal catalytic domains at the prongs of the "V" fork (Figs. 3A–C for FkpA, DsbC, and DsbG, respectively). The major difference between DsbC and DsbG is that the latter is larger, with an extended α-helical connection region altering the relative orientations of the C-terminal domains. For both proteins, the substrate-binding sites are thought to reside in hydrophobic patches located on the inside of the C-terminal domains and linkers, with the distance between the two binding sites able to change due to flexibility in the hinge region. However, in DsbG there are some additional acidic, negatively charged residues within the substrate-binding region, perhaps conferring a different substrate specificity to that of DsbC (Heras et al., 2004).

Despite the strong similarities in overall shape between the Dsb enzymes and FkpA, there is no discernable structural homology between them. The dimerization domains of both DsbC and DsbG are broadly similar and consist of a central β-sheet flanked by two α-helices, but they are both entirely different to the entwined α-helices of FkpA. In addition, the C-terminal, catalytic domains of DsbC and DsbG adopt thioredoxin folds typical of disulfide isomerases, while the equivalent domain of FkpA is an FKBP homologue (Heras et al., 2004). Therefore, the analogous molecular shapes could indicate convergent evolution: perhaps the flexible dimer is a particularly effective way of providing low affinity interactions with a broad variety of different substrates. Indeed, it has been observed that a clamp-like structure is a common feature of many chaperones, both ATP-dependent and ATP-independent (Stirling et al., 2006).

G. HdeA/B, the Periplasmic Acid Response Operon

The potential for the presence of additional, undiscovered periplasmic chaperones is highlighted by the recent characterization of two chaperones involved in survival of Gram-negative bacteria at low pH. Acid stress is one of the more common environmental perturbances of enteropathogenic bacteria, as the mammalian digestive tract exposes pathogens to pH levels as low as 1–3. In the absence of a protective response, such an acidic environment would cause rapid precipitation of many proteins leading to cell death. The periplasm in particular is highly susceptible to external pH changes as it is directly exposed to the external environment via the porins in the outer membrane. Two periplasmic gene products have been shown to be involved in the rapid acid response in enteropathogenic bacteria, HdeA (Gajiwala and Burley, 2000; Hong et al., 2005) and HdeB

(Kern *et al.*, 2007). These proteins are specifically protective at low pH (3 or below) and have been shown to function as periplasmic chaperones, binding to a wide variety of denatured proteins and preventing them from aggregating. HdeA is by far the best characterized of the two proteins; however, both HdeA and HdeB are thought to have similar structures and mechanisms of action. The main difference seems to be that HdeA is more effective for protecting cells at pH 2, while HdeB is more effective at pH 3, which suggests that the two proteins act under overlapping but distinct conditions (Kern *et al.*, 2007). HdeA and HdeB together give the highest cell survival rate *in vivo* under conditions of acid stress, and the two proteins together are necessary and sufficient to prevent low pH-induced aggregation of periplasmic extracts (Kern *et al.*, 2007; Malki *et al.*, 2008).

The initial crystal structures of HdeA showed it to be a dimer at physiological pH, with each monomer comprising a small, compact bundle of four α-helices, and the two units linked together by an extensive hydrophobic interface (Fig. 3D; Gajiwala and Burley, 2000; Kern *et al.*, 2007; Yang *et al.*, 1998). Gajiwala *et al.* were also able to show that this dimer association breaks at low pH, suggesting that monomerization was linked to chaperone activation; however, the mechanism for this was unclear at the time (Gajiwala and Burley, 2000). Subsequent studies confirmed the above result and showed that HdeA undergoes large and extremely rapid structural changes upon exposure to acid conditions. At pH values below ∼3.1–2.5, the dimer falls apart and each monomer partially unravels, exposing large areas of hydrophobic surface able to bind to hydrophobic patches exposed by acid-induced unfolding of substrates (Hong *et al.*, 2005; Tapley *et al.*, 2009). In this capacity, HdeA functions almost like a protein detergent, covering up water-insoluble surfaces and exposing a charged surface to the aqueous periplasmic environment. The rapid switching from a compact, inactive form under normal circumstances to an extended, functional state upon exposure to the stress conditions it is designed to protect against makes HdeA a highly efficient chaperone for its specific role (Tapley *et al.*, 2009).

III. CARRIER CHAPERONES

Although most proteins reach their final, stable conformation upon initial folding, for example, as they are transported through the Sec translocon into the periplasm, a small subset do not and are instead captured and maintained in higher energy, metastable states. Within the periplasm, this strategy is used by several systems to transduce energy from the cytoplasm to the outer membrane. Specific steric chaperones, which belong to the more general group of carrier chaperones, are responsible for capturing such proteins in their intermediate states and holding them until the energy can be used at a later point (Pauwels *et al.*, 2007). An example of this would be the proposed function of SurA, that is, carrying outer membrane proteins as unfolded polypeptides so as to preserve their folding energy for membrane insertion. Indeed, it is mainly for historical reasons that we have assigned SurA to the general chaperone category in this chapter. In addition to energy transduction, the activities of carrier chaperones can fulfil several different, more general requirements: they can protect their cargo from external factors such as proteases; they can render their substrates soluble in the aqueous environment of the periplasm; and they can target bound substrates to specific subcellular locations. The chaperones described in this section bind to specific substrates and carry out at least one and often more than one of these functions.

The scope of this chapter is limited to periplasmic chaperones that carry proteins. However, it should be noted that other periplasmic factors also use carrier chaperones for storage and transport, although these proteins are largely less well characterized. For example, metal ions are often sequestered in specific metallochaperones, which store metal ions for use as cofactors and prevent the toxicity that could be caused by their free presence in the periplasm (Bagai *et al.*, 2008; Hantke, 2005). In addition, one of the most plausible current models for the localization of LPS to the outer leaflet of the outer membrane invokes a potential chaperone (LptA), which would envelop the hydrophobic lipid group during transit through the periplasmic space (Bos *et al.*, 2007; Tran *et al.*, 2008). It has even been reported that some of the small molecule carrier proteins in the periplasm possess more general chaperone activity toward unfolded proteins, although such activity remains to be thoroughly explored (Richarme and Caldas, 1997).

A. The Chaperone/Usher Pathway

Gram-negative bacteria express a variety of different classes of adhesive surface organelles that allow them to bind specifically to host tissues and infect them. Among these, chaperone–usher (CU) pili constitute the

most abundant group of bacterial cell surface appendages and have
been the most extensively studied (Fronzes *et al.*, 2008; Nuccio and
Baumler, 2007). These adhesive structures consist of a series of different pilus subunits linked together into a fiber, the assembly of which
requires a soluble chaperone and an outer membrane assembly platform called an usher (Sauer *et al.*, 1999, 2000, 2004; Thanassi *et al.*,
1998). The periplasmic chaperone binds to the pilus subunits emerging
from the general Sec machinery and accomplishes three distinct functions: (1) it catalyzes the folding of the nascent pilus subunits (Barnhart
et al., 2000; Vetsch *et al.*, 2004), (2) it protects the pilus subunits from
degradation and nonproductive polymerization in the periplasm
(Hultgren *et al.*, 1989; Jones *et al.*, 1997), and (3) it targets the fiber
subunits to the usher (Dodson *et al.*, 1993; Thanassi *et al.*, 1998). Once
targeted to the outer membrane usher pore, the pilus subunits are
released from the chaperone and polymerized to form linear fibers at
the bacterial surface.

Depending on the adhesive organelle type and conserved structural
features in the chaperones, CU systems can be classified into two distinct
subfamilies: the FGL subfamily (with a long loop between β-strands
F_1–G_1 of the chaperone), which is involved in nonpilus or capsule-like
adhesive organelle assembly, and the FGS subfamily (short loop between
β-strands F_1–G_1), which is implicated in adhesive pilus assembly (Hung
et al., 1996; Zavialov *et al.*, 2007). Several structures of chaperone–subunit
complexes from each subfamily are available, including the archetypal
PapD chaperone from the *E. coli* P pilus (Holmgren and Branden, 1989;
Sauer *et al.*, 1999, 2002) and the FimC chaperone from the *E. coli* type 1
pilus (Choudhury *et al.*, 1999), both of which belong to the FGS subfamily; the Caf1M chaperone involved in the *Yersinia pestis* F1-antigen fiber
formation (Zavialov *et al.*, 2003) and the SafB chaperone from *Salmonella
typhimurium* atypical fimbriae (Remaut *et al.*, 2006) which represent the
FGL subfamily.

FGS and FGL chaperones present a similar fold: a boomerang- or
L-shaped structure of ~25 kDa, consisting of two immunoglobulin
(Ig)-like fold domains each comprising seven β-strands (strands A–G)
(see Fig. 4A). The pilus subunits also possess an Ig-like fold; however,
for the subunits this structure lacks the final, C-terminal β-strand (Fig. 4A;
Choudhury *et al.*, 1999; Sauer *et al.*, 1999). As a result, the subunit surface
exhibits a deep hydrophobic groove into which the chaperone binds by

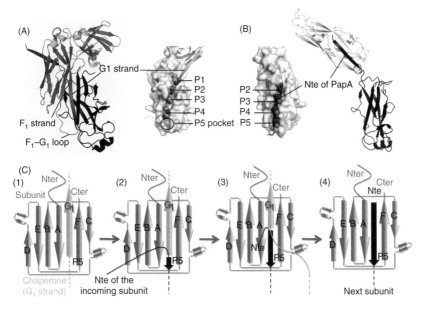

FIG. 4. The chaperone/usher pathway. (A) Two different views of the archetypal chaperone PapD in complex with the pilus rod subunit PapA [PDB code 2UY6 (Verger *et al.*, 2007)]. Left panel: cartoon representation of a PapA subunit (colored black) in donor-strand complementation (DSC) with the PapD chaperone (colored by secondary structure with α-helices in cyan, β-sheets in magenta, and turn regions in pink). The β-strands F_1 and G_1 and the F_1–G_1 loop between them are indicated. Right panel: surface representation of the PapA subunit with the G_1 strand of the chaperone shown in cartoon representation. The chaperone's residues P1–P4 bound within the P1–P4 sites/regions of the subunit's hydrophobic binding groove are highlighted in space-filling representation (carbon in green, oxygen in red, and nitrogen in blue). (B) Two different views of two pilus subunits in donor-strand exchange (DSE) [PDB code 2UY6 (Verger *et al.*, 2007)]. For clarity, the chaperone associated with the first subunit is not shown. Right panel: cartoon representation with one subunit in white and the other in black. Left panel: surface representation of the subunit in white at right, with the Nte of the subunit in black at right shown as a cartoon. Nte's residues P2–P5 bound within the P2–P5 sites/regions of the hydrophobic binding groove are highlighted in space-filling representation (carbon in black, oxygen in red, and nitrogen in blue). (C) Schematic diagram of the DSC and DSE mechanisms. (1) Topology diagram of the pilus subunit (in gray) showing the Ig-like fold complemented by the insertion of the G_1 strand from the chaperone (in green). In this donor-strand-complemented conformation, the pocket P5 is unoccupied. (2) Initial positioning of the incoming subunit's N-terminal extension (Nte, in black) via insertion of the P5 residue of the incoming subunit's Nte into the P5 pocket of the receiving subunit's groove. (3) Progressive insertion of the incoming subunit's Nte inside the receiving subunit's groove. While the Nte zips in, the chaperone's G1 strand zips out. (4) Completed DSE reaction. The chaperone is now completely released, and the receiving pilus subunit is stabilized by the completed insertion of the Nte from the incoming subunit. (See Color Insert.)

inserting the seventh β-strand of its N-terminal domain, the G_1 strand, thereby providing *in trans* the missing secondary structure. This mechanism by which the chaperone stabilizes pilus subunits is termed "donor-strand complementation" (DSC) (Sauer *et al.*, 1999). The G_1 strand is characterized by a conserved motif of four alternating hydrophobic residues, called P1–P4, which bind into the corresponding P1–P4 regions or pockets located within the subunit groove. Two critical, conserved, basic residues located between the two Ig-like domains of the chaperone also contribute to the interaction surface, anchoring the C-terminal carboxyl group of the subunit in place (Kuehn *et al.*, 1993; Slonim *et al.*, 1992). Through the DSC interaction, the chaperone completes the Ig-like fold of the subunit in a noncanonical manner, as the complementary G_1 strand from the chaperone runs parallel to the F_1 strand of the pilus subunit, rather than antiparallel as would be observed in canonical Ig-like folds (Fig. 4A; Barnhart *et al.*, 2000).

The presence of chaperone bound to the subunit prevents premature subunit polymerization by occupying the same hydrophobic groove as is required for the subunit–subunit interaction (Fig. 4B; Bullitt *et al.*, 1996; Kuehn *et al.*, 1991; Verger *et al.*, 2007). The subunit is indeed incorporated into the growing pilus via the "donor-strand exchange" (or DSE) reaction in which the G_1 strand of the chaperone is replaced by a sequence formed by the N-terminus of the incoming subunit (also termed "the N-terminal extension" or "Nte"; see Fig. 4B). Nte sequences contain a motif of alternating hydrophobic residues: three of these residues, also termed P2–P4, occupy the P2–P4 regions/pockets of the receiving subunit's groove once DSE is complete, replacing the P2–P4 residues of the chaperone. However, Ntes also contain another residue termed the "P5 residue" that occupies another region/pocket of the receiving subunit's groove, the P5 pocket, after DSE (shown in Fig. 4B). The DSE reaction is thought to proceed via a zip-in-zip-out mechanism (Remaut *et al.*, 2006), whereby the Nte of the incoming subunit initiates DSE by inserting its P5 residue into the P5 pocket of the receiving subunit's groove (Fig. 4C). This initial binding event is then followed by the zippering-in of the incoming subunit while the G_1 strand of the chaperone zips out in a concerted manner, resulting in the occupation of the P4–P2 regions of the receiving subunit's groove. The chaperone is then released to pick up another nascent subunit. This zippering mechanism was first demonstrated by using real-time, nondenaturing mass spectrometry to study *in vitro* DSE reactions. These experiments identified a transient ternary complex formed by the chaperone, the subunit, and a peptide mimicking the Nte of the next subunit (Remaut *et al.*, 2006). The formation of this ternary complex was shown to depend on the accessibility of the P5 pocket in the hydrophobic groove of the accepting pilus subunit. In addition, single-site alanine mutations of the P5, P4, and P3 residues in the Nte revealed a decreasing gradient of importance for these residues in the efficiency of DSE, lending further support to the zippering model (Remaut *et al.*, 2006).

From the description above, it is clear that the chaperones of the chaperone/usher pathway perform many of the functions required of periplasmic carrier proteins. First, they are important for subunit folding: fimbrial subunits are able to fold in the absence of the chaperone at low ionic and protein concentrations (Vetsch *et al.*, 2002); however, the presence of the chaperone accelerates the rate of folding 100-fold, bringing it up to physiologically relevant speeds (Vetsch *et al.*, 2004). In addition to catalyzing folding, the presence of the chaperone prevents pilus subunits from being degraded by proteases within the periplasm, blocks their spontaneous, unregulated polymerization, and targets them to the usher in the outer membrane. Finally, and particularly characteristic of steric chaperones, PapD-like chaperones are thought to act as a storage device for protein-folding energy. A crystal structure of a ternary complex composed of the chaperone with two subunits in DSE has revealed two different subunit conformations: one in which the subunit is "complemented" by the chaperone (i.e., before DSE; Fig. 4A) and the other in which it is "exchanged" by the Nte of the next subunit (after DSE; Fig. 4B; Verger *et al.*, 2007; Zavialov *et al.*, 2003). In the DSC state, bulky hydrophobic residues from the chaperone's donor strand intercalate between the two sheets of the subunit β-sandwich fold. After DSE, the subunit presents a more packed conformation since the Nte residues from the donor subunit are smaller. Based on this structural difference and the results of thermodynamic studies (Zavialov *et al.*, 2005), it has been proposed that the chaperone preserves the subunit folding energy by maintaining the subunit in a semifolded state prior to the DSE process. Thus, the incoming subunit would transit from a high-energy, semiunfolded state in the chaperone called the open state to a low-energy, folded state called the closed state (Sauer *et al.*, 2002; Zavialov *et al.*, 2003). The energetic difference between the two states would release sufficient free energy to drive the fiber assembly process, highlighting a role for the chaperone as an alternative cellular energy source to ATP hydrolysis or the proton-motive force.

B. *Protease Propeptides: Intramolecular Chaperones*

Most proteases secreted by prokaryotic cells are synthesized as inactive proenzymes. These precursors often have a prodomain, generally called propeptide or prosequence, that is covalently attached to the N-terminal or C-terminal end of the mature enzyme sequence (Wandersman, 1989). Propeptides perform a dual function: promoting the correct folding of the protease domain and blocking its proteolytic activity while within the cell, prior to delivery to the secretion machinery. Because this function exactly corresponds to the role of chaperones, such propeptides are often termed intramolecular chaperones or IMCs (Chen and Inouye, 2008; Inouye, 1991; McIver *et al.*, 1995; Shinde and Inouye, 1993). IMCs help their protein targets to overcome transition state energy barriers without using energy sources such as ATP hydrolysis. They are subsequently degraded either autocatalytically or by preexisting active enzyme leaving behind the mature folded protease, which is then exported to the extracellular medium (Inouye, 1991; Shinde and Inouye, 1993).

IMCs can be classified into two categories: type 1 (mostly Ntes), which encompasses the chaperones that assist with tertiary structure folding, and type 2 IMCs (mostly C-terminal extensions), which corresponds to chaperones that guide the formation of quaternary structure (Chen and Inouye, 2008). One of the best-studied examples of propeptide-mediated folding among Gram-negative bacteria is the α-lytic protease or αLP (Cunningham *et al.*, 1999), a digestive enzyme secreted by *Lysobacter enzymogenes* to degrade other soil microorganisms (Brayer *et al.*, 1979; Olson *et al.*, 1970). The αLP precursor initially synthesized consists of a signal peptide of 24 residues (essential for the secretion across the cytoplasmic membrane), a type 1 propeptide of 175 residues and a catalytic domain of 198 residues (Epstein and Wensink, 1988). The propeptide is autocatalytically cleaved within the periplasm, leading to a noncovalent αLP–propeptide complex. At this stage, the propeptide remains bound and maintains the protease in an inactive form until it is secreted into the extracellular milieu.

Because αLP has evolved to function in harsh environments, the mature, secreted form is highly resistant to degradation, with an extremely high barrier to unfolding of 26 kcal/mol — effectively preventing unfolding events that would render αLP susceptible to proteolysis (Sohl *et al.*, 1998). It has been shown that αLP without its propeptide folds to a stable but inactive molten-globule intermediate, which converts to native protease on an exceedingly slow timescale with a kinetic rate constant for spontaneous folding at 4 °C corresponding to a half-life of 1,800 years. The N-terminal propeptide accelerates αLP folding remarkably, such that its half-life of folding is now 23 s at 4 °C (Baker *et al.*, 1992; Sohl *et al.*, 1998).

Structure determination of the Gram-negative bacteria *L. enzymogenes* αLP in complex with its propeptide (or prodomain) was possible, thanks to a mutant protease variant with lower than normal activity, and addition of a protease inhibitor that prevented degradation of the propeptide (Sauter *et al.*, 1998). The propeptide is an α/β protein made up of two globular N-and C-domains connected by a rigid linker to form a C-shaped molecule. A pair of three-stranded β-sheets lines the concave surface while the convex surface consists of α-helices packed against these β-sheets (Fig. 5A). αLP is also made up of two distinct globular N- and C-terminal domains. The propeptide surrounds the C-terminal domain of αLP with an extensive buried interface area >4,000 Å^2 — much more than average protein–protein surface area of 1,600 Å^2 (Wodak and Janin, 2002) — and blocks the active site through the insertion of its C-terminal extremity in a substrate-like manner. The extensive capping of the chaperone allows a tight interaction within the propeptide–αLP complex, characterized by a remarkably high affinity ($K_D = 0.3$ nM; Peters *et al.*, 1998). This particular tight interaction has been shown to be responsible for the displacement of the thermodynamical equilibrium in favor of the folded αLP by lowering the energetic barrier between the native and the partially folded states (Sohl *et al.*, 1998).

The propeptide chaperone folds independently from the protease domain, which itself adopts a molten-globule folding intermediate state. Based on structural and mutagenesis studies, a sequential and cooperative model of αLP folding has been proposed: the propeptide first stabilizes the C-terminal domain of the mature protease via a five-stranded β-sheet interaction, formed by three strands from the propeptide and a conserved β-hairpin from the C-terminal domain of the protease. In this

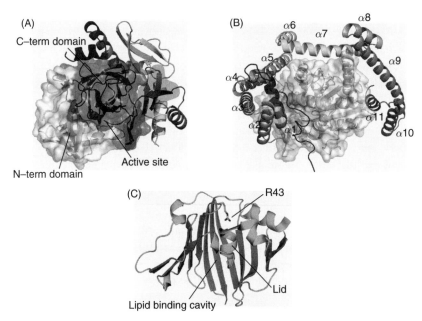

FIG. 5.   Structures of carrier chaperones. (A) Crystal structure of α-lytic protease (αLP) with its prodomain chaperone [PDB code 4PRO (Sauter *et al.*, 1998)]. The prodomain is in cartoon representation colored in rainbow and the αLP is in both cartoon/surface representations. The N- and C-terminal domains of αLP are colored, respectively, in light gray and black. The location of the active site of LP is indicated. It is occupied by the insertion of the prodomain C-terminal tail in red. (B) Crystal structure of lipase A (LipA) bound to its chaperone Lif [PDB code 2ES4 (Pauwels *et al.*, 2006)]. The chaperone is in cartoon representation and colored in rainbow. LipA is in cartoon and surface representations, colored in gray. (C) Structure of LolA [PDB code 1IWL (Takeda *et al.*, 2003)] shown in cartoon representation with α-helical regions in cyan, β-sheet in magenta, and loop regions in pink. Arginine 43 is marked out in stick representation with nitrogen atoms colored in blue. (See Color Insert.)

conformation, the C-terminal tail of the chaperone blocks the nascent active site of the protease. This interaction would then allow structural rearrangement of the N-terminal domain of the protease, which would pack tightly against the N-terminal domain of the chaperone to produce the mature conformation (Anderson *et al.*, 1999; Cunningham and Agard, 2003; Cunningham *et al.*, 2002; Sauter *et al.*, 1998; Truhlar and Agard, 2005).

By helping its cognate protease to fold into a kinetically trapped metastable state, the αLP chaperone neatly sidesteps the problem of producing a very stable structure without an external energy supply. It should be noted that IMCs are not exclusively found in the periplasm of Gram-negative bacteria, as the first secreted protease reported to require a prosequence for folding was subtilisin from the Gram-positive species *Bacillus subtilis* (Ikemura *et al.*, 1987; Inouye, 1991). However, it is perhaps not so surprising that this particular solution to the periplasm energy barrier problem should be replicated even in species without a periplasm. The fact that the chaperone and substrate are covalently linked allows the use of a chaperone even when there is no mechanism to retain it at the site of protein export.

C. Lif, an Enzyme Carrier Protein

Gram-negative bacteria secrete a variety of hydrolytic enzymes as virulence factors via the type II secretion machinery (Filloux, 2004), including the bacterial lipases, which catalyze the hydrolysis of triglyceride ester bonds to free fatty acids and glycerol. Some of these need a so-called lipase-specific foldase (Lif), a steric chaperone unusually anchored to the inner membrane via a N-terminal transmembrane segment. This transmembrane region is not mandatory for the chaperone function but is probably there to ensure that the chaperone is retained within the cell while the lipase is secreted (El Khattabi *et al.*, 1999). Lif is thought to promote the native folding of the lipase once in the periplasm before extracellular secretion (Jaeger *et al.*, 1994): in the absence of Lif, the lipase is degraded or accumulates as inactive aggregates in the periplasm (Frenken *et al.*, 1993). The molecular mechanism of Lif function is still unknown, but unlike the protease steric chaperones, Lif is neither encoded within the primary sequence of its target protein, nor does it show an inhibitory activity toward its lipase target. Preliminary *in vitro* refolding experiments have shown that the Gram-negative *Burkholderia glumae* lipase A (LipA) without its cognate chaperone is able to adopt a fold similar to the native conformation, but more compact and inactive. Thus, the Lif chaperone would help the lipase to overcome an energetic barrier late in its folding pathway rather than rescuing it from a molten-globule state as was observed for proteases (El Khattabi *et al.*, 2000).

A crystal structure of LipA in complex with its Lif is the first example of a structure of a member of this peculiar class of nonintramolecular steric chaperones (Pauwels *et al.*, 2006). Lif embraces the lipase through an extended α-helical structure consisting of 11 α-helices (Fig. 5B), in which two minidomains (α1–α3 and α9–α11) are connected by a long α-helical motif (α4–α8). Sequence alignment and mutagenesis have shown that Lifs present a conserved and functionally critical motif RxxFDY(F/C)L(S/T)A (where x is any residue; Rosenau *et al.*, 2004; Shibata *et al.*, 1998) localized on helix α1 of the N-terminal domain. The high flexibility of Lif compared to its target LipA (the average B-factor of Lif is twice that of LipA in the crystal structure) may reflect the role of the chaperone as a dynamic platform during the lipase folding. The interface between Lif and its cognate lipase is remarkably large, with $5{,}400\,\text{Å}^2$ of buried solvent-accessible surface. This is in accordance with the very tight interaction measured by plasmon surface resonance ($K_D = 5\,\text{nM}$; Pauwels *et al.*, 2006). Since Lif is not degradated upon lipase folding in the native state, the type II secretion machinery probably accomplishes the costly energetic dissociation between the chaperone and the lipase.

It is still unclear why lipases have evolved to require a chaperone and interact with it using such an extensive surface area. Indeed, some homologous lipases, notably the cold-active lipase from the psychrophile *Pseudomonas fragi*, seem to be able to fold without the need for any chaperone (Alquati *et al.*, 2002). This cold-active lipase has much higher content of surface-exposed charged residues than the Lif-dependent lipases, and the protein core is probably less compact and hydrophobic. Thus, it seems that Lif sterically traps the lipase in a slightly more flexible, less stable but active conformation, with a highly associated energetic cost. Perhaps, as with the proteases, this difference reflects the varying stresses encountered by the lipase after secretion, with the use of a chaperone yielding a less easily unfolded enzyme at the added expense of synthesizing an additional protein. Structural and thermodynamical comparison studies between Lif-dependent and Lif-independent lipases, as well as structural characterization of trapped folding intermediates, would throw light on chaperone function from a molecular point of view.

D. *Lipoprotein Chaperoning by LolA*

Many bacterial proteins are tethered to the inner or outer membranes via long acyl chains. These lipid anchors are covalently attached to specific N-terminal cysteines of periplasmic proteins as they emerge from the Sec machinery, forming lipoproteins, and can then be inserted into the appropriate membrane. The highly hydrophobic tails of lipoproteins prevent them from crossing the periplasmic space alone, and a specialized Lol pathway is used to sort and traffic those destined for the outer membrane. The Lol machinery comprises five proteins, all of which are essential for function: the inner membrane complex LolCDE, the periplasmic chaperone LolA, and the outer membrane-associated lipoprotein LolB. Outer membrane lipoproteins are loaded onto LolA by the LolCDE complex in an ATP-dependent manner, cross the periplasm with the help of the chaperone, and are then passed to LolB for membrane insertion (Takeda et al., 2003).

Crystallographic data for the chaperone LolA (Oguchi et al., 2008; Takeda et al., 2003) show an 11-strand antiparallel β-sheet, folded over itself to give a β-barrel-like structure with one side open (Fig. 5C). The open face of the β-barrel is covered by three α-helices which together form a lid, held in place by a salt bridge between Arg 43 on the loop between strands β2 and β3 and the backbone of helices 1 and 2. The inner surfaces of the β-barrel and α-helices together form a highly hydrophobic cavity, which is protected from solvent access by several conserved aromatic residues surrounding the entrance. This hydrophobic core contains the binding site for the acyl moieties of lipoproteins, sequestering them away from the aqueous environment during transit between the inner and outer membranes (Nakada et al., 2009). Structural studies using an R43L mutant of LolA showed that the protein can adopt two distinct conformations: an open form in which the lid is swung away from the centre of the barrel allowing lipids to access the cavity and a closed form in which the lid blocks the gap. Only the closed form is observed for the native protein, suggesting that it is the more stable state; however, the open form is thought to resemble the conformation in the presence of a lipoprotein ligand (Oguchi et al., 2008).

Once LolA has bound a lipoprotein, it moves to the outer membrane and rapidly releases its substrate to the membrane insertion protein LolB. This transfer is highly efficient: under normal physiological conditions, it is not possible to detect LolA:lipoprotein complexes in the periplasm. Despite a lack of sequence homology, LolB adopts a very similar fold to LolA, forming an open β-barrel lined with hydrophobic residues into which lipid anchors can bind (Takeda et al., 2003). The main difference between the structures of LolA and LolB is the absence of a lid on LolB, leaving the lipid-binding cavity permanently open and solvent accessible. The absence of a lid domain could explain the higher affinity of LolB for lipoproteins. This difference in affinity is important: there is no ATP in the periplasm to drive insertion of lipoproteins into the membrane, so the transfer from LolA to LolB and from LolB to the membrane must be energetically favorable to ensure efficient transit of substrates through the Lol pathway. Indeed, the R43L mutant of LolA — in which the stability of the open form is increased relative to the closed form — is defective for the outer membrane localization of lipoproteins, with LolA:lipoprotein complexes accumulating in the periplasm (Oguchi *et al.*, 2008).

Recent results show how the interaction between LolA and LolB might facilitate transfer of the acyl moiety between the two proteins. NMR distance constraints suggest that the inner surface of one end of the LolA barrel contacts the outer surface of one end of the LolB barrel, bringing the two substrate-binding grooves together to form a long hydrophobic channel (Nakada *et al.*, 2009). The lipid anchor could then slide directly from LolA to LolB, perhaps pushed by the closing of the LolA lid. Thus, LolA is exquisitely evolved to chaperone highly specific substrates through the periplasm. ATP hydrolysis at the inner membrane loads N-terminally lipidated proteins into the central cavity of the chaperone, where they are kept protected from the aqueous environment of the periplasm. However, this configuration is only metastable: the substrate-binding energy is retained and used to pass the lipoprotein substrate efficiently to LolB for membrane insertion.

IV. CONCLUSION

The periplasm of Gram-negative bacteria provides a unique and challenging environment for protein folding and stabilization in that it has no ATP and is highly exposed to fluctuations in the external environment. The lack of energy source provides a particular barrier to processes that require the input of energy, for example, biosynthesis at the outer membrane. To overcome both these problems, organisms such as *E. coli* employ a number of general and specialized chaperones that assist with the normal function of the periplasm and also alleviate the effects of environmental stress. In this chapter, we have classified the chaperones of the periplasm into two categories based on their functional properties: the folding chaperones, which perform an analogous function to the classical heat shock proteins of the cytoplasm such as GroEL/GroES and DnaK, and the carrier chaperones, which are involved in stabilization and transport of specific substrates. However, it becomes apparent upon attempting to assign chaperones such as SurA or PapD that there is substantial overlap between the two classes. Indeed, it is perhaps only the small dimeric chaperones such as FkpA and DsbC that are entirely limited to one of the two activities.

The normal functioning of the periplasm relies on both carrier chaperones and folding chaperones, as shown in Fig. 6. For correct export of the bulk of integral outer membrane proteins, which by necessity expose large patches of lipophilic surface area, there is an absolute requirement for at least one of the two chaperones Skp or SurA. According to the model displayed in Fig. 6, these two chaperones are able to pick up outer membrane proteins as they emerge in the unfolded state from the Sec translocon and carry them to the BAM complex in the outer membrane, which then catalyzes their membrane insertion. In this case, the two chaperones would essentially be functioning as outer membane protein carrier proteins. It should be noted there are also reports of some outer membrane proteins crossing the periplasm in a relatively folded state (Brandon and Goldberg, 2001; Krojer et al., 2008) and this might represent an alternate route for outer membrane biogenesis. Indeed, the pore-forming secretins of many bacterial secretion systems are not reported to use the standard Skp/SurA pathway and are instead thought to localize to the outer membrane with the aid of specific "pilotins," whereupon they self-assemble into their final multimeric structures (Okon et al., 2008).

Lipoproteins also have a specific chaperone requirement, as their hydrophobic lipid anchors are unable to traverse the periplasm without the aid of LolA. The LolCDE complex in the inner membrane consumes ATP to load lipoproteins onto LolA, which shuttles them across the periplasm and hands them over to LolB for membrane insertion (Fig. 6). The chaperones of the chaperone/usher pathway carry out a similar role in pilus biogenesis, taking up pilus subunits and ferrying them to the pilus assembly site, the usher. However, their involvement in the correct folding of the pilus subunits further blurs the distinction between folding and transport.

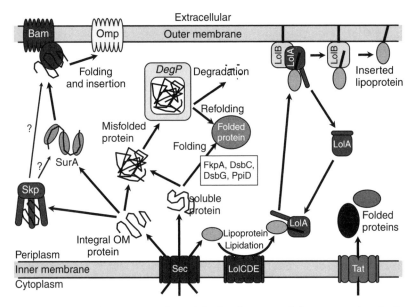

FIG. 6.   Schematic representation of the involvement of chaperones (marked out in italics) in protein folding in the periplasm. Proteins are exported to the periplasm via the Sec machinery, which transports proteins in the unfolded state, or via the Tat machinery, which translocates folded proteins. Unfolded outer membrane proteins are picked up by Skp or SurA and are then transported to the Bam complex in the outer membrane for insertion. Soluble proteins fold either directly or with the aid of periplasmic folding factors such as FkpA. Lipoproteins are lipidated by the LolCDE complex in the inner membrane, loaded onto LolA, and then transported to LolB in the outer membrane for insertion. Any of these proteins can potentially misfold, in which case they are either rescued and refolded by DegP or other general chaperones, or are degraded to prevent them damaging the integrity of the cell.

One of the key features of carrier chaperones identified so far is the ability to compensate for the lack of ATP in the periplasm. Such energetic functions, uniquely necessitated by the specific environment in which their substrates reside, can be divided into two categories: energy transduction, in which the energy is stored to power assembly steps at the outer membrane, and kinetic trap mechanisms, where secreted proteins are folded into conformations from which they are unable to unfold once the chaperone is removed. These two distinct functions are helpful for understanding the roles of periplasmic chaperones; however, they are by no means exclusive: the chaperone/usher pathway uses both mechanisms to ensure rapid, efficient production of pili that are highly stable once displayed on the cell surface.

Energy transduction systems in the periplasm include the insertion pathways for membrane proteins and the chaperone/usher pilus biogenesis pathway. For example, SurA and Skp are thought to store the folding energy of outer membrane proteins, assisting with their insertion into the outer membrane. Similarly, PapD only partially folds its cognate pilus subunits, storing the remainder of the folding energy to drive the DSE reaction. And the loading of lipoprotein onto LolA makes use of ATP hydrolysis at the inner membrane, using this energy to push its substate to LolB. Bacterial secretion systems also require energy transduction systems to power export of their substrates through the outer membrane; however, the majority of these are large complexes spanning the entire cell envelope and are thought to use ATP energy directly (Durand et al., 2009). The exception to this is the type V secretion systems — autotransporters and two-partner secretion systems — which use a pore-forming translocator domain to secrete a functional passenger domain. The detailed secretion processes of these systems have not yet been established, but the final export step is in many cases thought to be driven by folding of the passenger domain on the extracellular side of the outer membrane. This would imply a mechanism through which premature folding on the periplasmic side of the membrane is prevented — perhaps involving extrinsic periplasmic chaperones or a domain with IMC activity (Durand et al., 2009; Thanassi et al., 2005).

The kinetic trap mechanism of periplasmic chaperones, used by the protease propeptides, the chaperone–usher pathways and possibly the lipase folding factors, makes use of the ability of some chaperones to catalyze folding directly. By lowering the energy barrier to folding or assembly, they allow noncovalent reactions to occur on a timescale that is physiologically relevant. However, once their target proteins has been secreted or translocated to the cell surface, the chaperones are no longer present and so the reverse reaction is blocked by an effective kinetic barrier. This function of chaperones accounts for the high stability of pili involved in bacterial adhesion and of some secreted proteases.

Although the best-characterized chaperones in the periplasm have specific substrates and carrier chaperone functionality, a variety of more general chaperones and folding catalysts have been identified in the periplasm, including DegP, FkpA, DsbC, DsbG, and PpiD (Fig. 6). These could potentially be involved in the normal folding of any other type of protein, including membrane-bound, secreted, or soluble proteins. However, aside from the disulfide bond-forming activity of the Dsb enzymes and the proteolytic function of DegP, there is little evidence for their direct requirement under normal physiological conditions. Instead, a likely role for this group of proteins is in the protection against protein unfolding caused by extrinsic factors such as heat shock. Such environmental changes can cause native proteins to unfold, reducing their levels within the periplasm and exposing normally buried hydrophobic surfaces to the external environment. Under these circumstances, chaperones with otherwise poorly understood function might bind to these hydrophobic surfaces, preventing aggregation and allowing the partially unfolded proteins time to reach their properly folded states once again.

Very few soluble periplasmic proteins have actually been shown to require the presence of chaperones for their correct folding or stabilization under normal conditions. Indeed, most of the studies that identified the chaperone activity in the chaperones mentioned above have used model cytoplasmic substrates, rather than the periplasmic proteins they would encounter *in vivo*. It has been suggested that this lack of established chaperone targets in the periplasm could be explained by an inherent resistance of all periplasmic proteins to aggregation (Liu *et al.*, 2004). And although the detailed conclusions of the above paper are questionable, it is nonetheless plausible that periplasmic proteins could have evolved a propensity to fold properly without outside assistance and to remain stable under a variety of different conditions. However, as illustrated by the specific example of the acid shock response, there is a definite requirement for protection of proteins against folding stress, and it seems likely that the conserved chaperones within the periplasm would be responsible for this protection. It would be of interest to understand more fully the various roles of the general chaperone in protein folding and stabilization, and how their interplay keeps bacterial cells alive under varying conditions. Understanding these processes could not only lead to improved recombinant expression systems in bacteria (Mansell *et al.*, 2008; Schlapschy *et al.*, 2006), but also provide targets for antibacterial agents that act only under specific conditions.

REFERENCES

Alquati, C., De Gioia, L., Santarossa, G., Alberghina, L., Fantucci, P., and Lotti, M. (2002). The cold-active lipase of *Pseudomonas fragi*. Heterologous expression, biochemical characterization and molecular modeling. *Eur. J. Biochem.* **269**, 3321–3328.

Anderson, D. E., Peters, R. J., Wilk, B., and Agard, D. A. (1999). Alpha-lytic protease precursor: Characterization of a structured folding intermediate. *Biochemistry* **38**, 4728–4735.

Antonoaea, R., Furst, M., Nishiyama, K., and Muller, M. (2008). The periplasmic chaperone PpiD interacts with secretory proteins exiting from the SecYEG translocon. *Biochemistry* **47**, 5649–5656.

Arie, J. P., Sassoon, N., and Betton, J. M. (2001). Chaperone function of FkpA, a heat shock prolyl isomerase, in the periplasm of *Escherichia coli*. *Mol. Microbiol.* **39**, 199–210.

Bagai, I., Rensing, C., Blackburn, N. J., and McEvoy, M. M. (2008). Direct metal transfer between periplasmic proteins identifies a bacterial copper chaperone. *Biochemistry* **47**, 11408–11414.

Baker, D., Sohl, J. L., and Agard, D. A. (1992). A protein-folding reaction under kinetic control. *Nature* **356**, 263–265.

Barnhart, M. M., Pinkner, J. S., Soto, G. E., Sauer, F. G., Langermann, S., Waksman, G., Frieden, C., and Hultgren, S. J. (2000). PapD-like chaperones provide the missing information for folding of pilin proteins. *Proc. Natl. Acad. Sci. USA* **97**, 7709–7714.

Behrens, S., Maier, R., de Cock, H., Schmid, F. X., and Gross, C. A. (2001). The SurA periplasmic PPIase lacking its parvulin domains functions *in vivo* and has chaperone activity. *EMBO J.* **20**, 285–294.

Bitto, E., and McKay, D. B. (2002). Crystallographic structure of SurA, a molecular chaperone that facilitates folding of outer membrane porins. *Structure* **10**, 1489–1498.

Bitto, E., and McKay, D. B. (2003). The periplasmic molecular chaperone protein SurA binds a peptide motif that is characteristic of integral outer membrane proteins. *J. Biol. Chem.* **278**, 49316–49322.

Bitto, E., and McKay, D. B. (2004). Binding of phage-display-selected peptides to the periplasmic chaperone protein SurA mimics binding of unfolded outer membrane proteins. *FEBS Lett.* **568**, 94–98.

Bos, M. P., Robert, V., and Tommassen, J. (2007). Biogenesis of the gram-negative bacterial outer membrane. *Annu. Rev. Microbiol.* **61**, 191–214.

Bothmann, H., and Pluckthun, A. (2000). The periplasmic *Escherichia coli* peptidylprolyl *cis,trans*-isomerase FkpA. I. Increased functional expression of antibody fragments with and without *cis*-prolines. *J. Biol. Chem.* **275**, 17100–17105.

Brandon, L. D., and Goldberg, M. B. (2001). Periplasmic transit and disulfide bond formation of the autotransported *Shigella* protein IcsA. *J. Bacteriol.* **183**, 951–958.

Brayer, G. D., Delbaere, L. T., and James, M. N. (1979). Molecular structure of the alpha-lytic protease from *Myxobacter* 495 at 2.8 Angstroms resolution. *J. Mol. Biol.* **131**, 743–775.

Bulieris, P. V., Behrens, S., Holst, O., and Kleinschmidt, J. H. (2003). Folding and insertion of the outer membrane protein OmpA is assisted by the chaperone Skp and by lipopolysaccharide. *J. Biol. Chem.* **278**, 9092–9099.

Bullitt, E., Jones, C. H., Striker, R., Soto, G., Jacob-Dubuisson, F., Pinkner, J., Wick, M. J., Makowski, L., and Hultgren, S. J. (1996). Development of pilus organelle subassemblies *in vitro* depends on chaperone uncapping of a beta zipper. *Proc. Natl. Acad. Sci. USA* **93**, 12890–12895.

CastilloKeller, M., and Misra, R. (2003). Protease-deficient DegP suppresses lethal effects of a mutant OmpC protein by its capture. *J. Bacteriol.* **185**, 148–154.

Chen, J., Song, J. L., Zhang, S., Wang, Y., Cui, D. F., and Wang, C. C. (1999). Chaperone activity of DsbC. *J. Biol. Chem.* **274**, 19601–19605.

Chen, R., and Henning, U. (1996). A periplasmic protein (Skp) of *Escherichia coli* selectively binds a class of outer membrane proteins. *Mol. Microbiol.* **19**, 1287–1294.

Chen, Y. J., and Inouye, M. (2008). The intramolecular chaperone-mediated protein folding. *Curr. Opin. Struct. Biol.* **18**, 765–770.

Choudhury, D., Thompson, A., Stojanoff, V., Langermann, S., Pinkner, J., Hultgren, S. J., and Knight, S. D. (1999). X-ray structure of the FimC-FimH chaperone-adhesin complex from uropathogenic *Escherichia coli*. *Science* **285**, 1061–1066.

Clausen, T., Southan, C., and Ehrmann, M. (2002). The HtrA family of proteases: implications for protein composition and cell fate. *Mol. Cell.* **10**, 443–455.

Collet, J. F., and Bardwell, J. C. (2002). Oxidative protein folding in bacteria. *Mol. Microbiol.* **44**, 1–8.

Cunningham, E. L., and Agard, D. A. (2003). Interdependent folding of the N- and C-terminal domains defines the cooperative folding of alpha-lytic protease. *Biochemistry* **42**, 13212–13219.

Cunningham, E. L., Jaswal, S. S., Sohl, J. L., and Agard, D. A. (1999). Kinetic stability as a mechanism for protease longevity. *Proc. Natl. Acad. Sci. USA* **96**, 11008–11014.

Cunningham, E. L., Mau, T., Truhlar, S. M., and Agard, D. A. (2002). The pro region N-terminal domain provides specific interactions required for catalysis of alpha-lytic protease folding. *Biochemistry* **41**, 8860–8867.

Dartigalongue, C., and Raina, S. (1998). A new heat-shock gene, ppiD, encodes a peptidyl-prolyl isomerase required for folding of outer membrane proteins in *Escherichia coli*. *EMBO J.* **17**, 3968–3980.

Dodson, K. W., Jacob-Dubuisson, F., Striker, R. T., and Hultgren, S. J. (1993). Outer-membrane PapC molecular usher discriminately recognizes periplasmic chaperone-pilus subunit complexes. *Proc. Natl. Acad. Sci. USA* **90**, 3670–3674.

Driessen, A. J., and Nouwen, N. (2008). Protein translocation across the bacterial cytoplasmic membrane. *Annu. Rev. Biochem.* **77**, 643–667.

Duguay, A. R., and Silhavy, T. J. (2004). Quality control in the bacterial periplasm. *Biochim. Biophys. Acta* **1694**, 121–134.

Durand, E., Verger, D., Rego, A. T., Chandran, V., Meng, G., Fronzes, R., and Waksman, G. (2009). Structural biology of bacterial secretion systems in gram-negative pathogens — potential for new drug targets. *Infect. Disord. Drug. Targets* **9**, 518–547.

El Khattabi, M., Ockhuijsen, C., Bitter, W., Jaeger, K. E., and Tommassen, J. (1999). Specificity of the lipase-specific foldases of gram-negative bacteria and the role of the membrane anchor. *Mol. Gen. Genet.* **261**, 770–776.

El Khattabi, M., Van Gelder, P., Bitter, W., and Tommassen, J. (2000). Role of the lipase-specific foldase of *Burkholderia glumae* as a steric chaperone. *J. Biol. Chem.* **275**, 26885–26891.

Epstein, D. M., and Wensink, P. C. (1988). The alpha-lytic protease gene of *Lysobacter enzymogenes*. The nucleotide sequence predicts a large prepro-peptide with homology to pro-peptides of other chymotrypsin-like enzymes. *J. Biol. Chem.* **263**, 16586–16590.

Filloux, A. (2004). The underlying mechanisms of type II protein secretion. *Biochem. Biophys. Acta.* **1694**, 163–179.

Frenken, L. G., de Groot, A., Tommassen, J., and Verrips, C. T. (1993). Role of the lipB gene product in the folding of the secreted lipase of *Pseudomonas glumae*. *Mol. Microbiol.* **9**, 591–599.

Fronzes, R., Remaut, H., and Waksman, G. (2008). Architectures and biogenesis of non-flagellar protein appendages in gram-negative bacteria. *EMBO J.* **27**, 2271–2280.

Gajiwala, K. S., and Burley, S. K. (2000). HDEA, a periplasmic protein that supports acid resistance in pathogenic enteric bacteria. *J. Mol. Biol.* **295**, 605–612.

Hantke, K. (2005). Bacterial zinc uptake and regulators. *Curr. Opin. Microbiol.* **8**, 196–202.

Harms, N., Koningstein, G., Dontje, W., Muller, M., Oudega, B., Luirink, J., and de Cock, H. (2001). The early interaction of the outer membrane protein phoe with the periplasmic chaperone Skp occurs at the cytoplasmic membrane. *J. Biol. Chem.* **276**, 18804–18811.

Hennecke, G., Nolte, J., Volkmer-Engert, R., Schneider-Mergener, J., and Behrens, S. (2005). The periplasmic chaperone SurA exploits two features characteristic of integral outer membrane proteins for selective substrate recognition. *J. Biol. Chem.* **280**, 23540–23548.

Heras, B., Edeling, M. A., Schirra, H. J., Raina, S., and Martin, J. L. (2004). Crystal structures of the DsbG disulfide isomerase reveal an unstable disulfide. *Proc. Natl. Acad. Sci. USA* **101**, 8876–8881.

Hodak, H., Wohlkonig, A., Smet-Nocca, C., Drobecq, H., Wieruszeski, J. M., Senechal, M., Landrieu, I., Locht, C., Jamin, M., and Jacob-Dubuisson, F. (2008). The peptidyl-prolyl isomerase and chaperone Par27 of *Bordetella pertussis* as the prototype for a new group of parvulins. *J. Mol. Biol.* **376**, 414–426.

Holmgren, A., and Branden, C. I. (1989). Crystal structure of chaperone protein PapD reveals an immunoglobulin fold. *Nature* **342**, 248–251.

Hong, W., Jiao, W., Hu, J., Zhang, J., Liu, C., Fu, X., Shen, D., Xia, B., and Chang, Z. (2005). Periplasmic protein HdeA exhibits chaperone-like activity exclusively within stomach pH range by transforming into disordered conformation. *J. Biol. Chem.* **280**, 27029–27034.

Hu, K., Galius, V., and Pervushin, K. (2006). Structural plasticity of peptidyl-prolyl isomerase sFkpA is a key to its chaperone function as revealed by solution NMR. *Biochemistry* **45**, 11983–11991.

Hultgren, S. J., Lindberg, F., Magnusson, G., Kihlberg, J., Tennent, J. M., and Normark, S. (1989). The PapG adhesin of uropathogenic *Escherichia coli* contains separate regions for receptor binding and for the incorporation into the pilus. *Proc. Natl. Acad. Sci. USA* **86**, 4357–4361.

Hung, D. L., Knight, S. D., Woods, R. M., Pinkner, J. S., and Hultgren, S. J. (1996). Molecular basis of two subfamilies of immunoglobulin-like chaperones. *EMBO J.* **15**, 3792–3805.

Ikemura, H., Takagi, H., and Inouye, M. (1987). Requirement of pro -sequence for the production of active subtilisin E in *Escherichia coli. J. Biol. Chem.* **262**, 7859–7864.

Inaba, K. (2009). Disulfide bond formation system in *Eschericia coli. J. Biochem.* doi:10.1093/jb/mvp102.

Inouye, M. (1991). Intramolecular chaperone: The role of the pro -peptide in protein folding. *Enzyme* **45**, 314–321.

Jaeger, K. E., Ransac, S., Dijkstra, B. W., Colson, C., van Heuvel, M., and Misset, O. (1994). Bacterial lipases. *FEMS Microbiol. Rev.* **15**, 29–63.

Jiang, J., Zhang, X., Chen, Y., Wu, Y., Zhou, Z. H., Chang, Z., and Sui, S. F. (2008). Activation of DegP chaperone-protease via formation of large cage-like oligomers upon binding to substrate proteins. *Proc. Natl. Acad. Sci. USA* **105**, 11939–11944.

Jomaa, A., Damjanovic, D., Leong, V., Ghirlando, R., Iwanczyk, J., and Ortega, J. (2007). The inner cavity of *Escherichia coli* DegP protein is not essential for molecular chaperone and proteolytic activity. *J. Bacteriol.* **189**, 706–716.

Jones, C. H., Danese, P. N., Pinkner, J. S., Silhavy, T. J., and Hultgren, S. J. (1997). The chaperone-assisted membrane release and folding pathway is sensed by two signal transduction systems. *EMBO J.* **16**, 6394–6406.

Justice, S. S., Hunstad, D. A., Harper, J. R., Duguay, A. R., Pinkner, J. S., Bann, J., Frieden, C., Silhavy, T. J., and Hultgren, S. J. (2005). Periplasmic peptidyl prolyl cis-trans isomerases are not essential for viability, but SurA is required for pilus biogenesis in *Escherichia coli. J. Bacteriol.* **187**, 7680–7686.

Kern, R., Malki, A., Abdallah, J., Tagourti, J., and Richarme, G. (2007). *Escherichia coli* HdeB is an acid stress chaperone. *J. Bacteriol.* **189**, 603–610.

Kleerebezem, M., Heutink, M., and Tommassen, J. (1995). Characterization of an *Escherichia coli* rotA mutant, affected in periplasmic peptidyl-prolyl *cis/trans* isomerase. *Mol. Microbiol.* **18**, 313–320.

Kleinschmidt, J. H., den Blaauwen, T., Driessen, A. J., and Tamm, L. K. (1999). Outer membrane protein A of Escherichia coli inserts and folds into lipid bilayers by a concerted mechanism. *Biochemistry* **38**, 5006–5016.

Knowles, T. J., Scott-Tucker, A., Overduin, M., and Henderson, I. R. (2009). Membrane protein architects: The role of the BAM complex in outer membrane protein assembly. *Nat. Rev. Microbiol.* **7**, 206–214.

Korndorfer, I. P., Dommel, M. K., and Skerra, A. (2004). Structure of the periplasmic chaperone Skp suggests functional similarity with cytosolic chaperones despite differing architecture. *Nat. Struct. Mol. Biol.* **11**, 1015–1020.

Krojer, T., Garrido-Franco, M., Huber, R., Ehrmann, M., and Clausen, T. (2002). Crystal structure of DegP (HtrA) reveals a new protease-chaperone machine. *Nature* **416**, 455–459.

Krojer, T., Sawa, J., Schafer, E., Saibil, H. R., Ehrmann, M., and Clausen, T. (2008). Structural basis for the regulated protease and chaperone function of DegP. *Nature* **453**, 885–890.

Kuehn, M. J., Normark, S., and Hultgren, S. J. (1991). Immunoglobulin-like PapD chaperone caps and uncaps interactive surfaces of nascently translocated pilus subunits. *Proc. Natl. Acad. Sci. USA* **88**, 10586–10590.

Kuehn, M. J., Ogg, D. J., Kihlberg, J., Slonim, L. N., Flemmer, K., Bergfors, T., and Hultgren, S. J. (1993). Structural basis of pilus subunit recognition by the PapD chaperone. *Science* **262**, 1234–1241.

Liu, Y., Fu, X., Shen, J., Zhang, H., Hong, W., and Chang, Z. (2004). Periplasmic proteins of *Escherichia coli* are highly resistant to aggregation: Reappraisal for roles of molecular chaperones in periplasm. *Biochem. Biophys. Res. Commun.* **316**, 795–801.

Malki, A., Le, H. T., Milles, S., Kern, R., Caldas, T., Abdallah, J., and Richarme, G. (2008). Solubilization of protein aggregates by the acid stress chaperones HdeA and HdeB. *J. Biol. Chem.* **283**, 13679–13687.

Mansell, T. J., Fisher, A. C., and DeLisa, M. P. (2008). Engineering the protein folding landscape in gram-negative bacteria. *Curr. Protein. Pept. Sci.* **9**, 138–149.

McCarthy, A. A., Haebel, P. W., Torronen, A., Rybin, V., Baker, E. N., and Metcalf, P. (2000). Crystal structure of the protein disulfide bond isomerase, DsbC, from *Escherichia coli*. *Nat. Struct. Biol.* **7**, 196–199.

McIver, K. S., Kessler, E., Olson, J. C., and Ohman, D. E. (1995). The elastase propeptide functions as an intramolecular chaperone required for elastase activity and secretion in *Pseudomonas aeruginosa*. *Mol. Microbiol.* **18**, 877–889.

Messens, J., and Collet, J. F. (2006). Pathways of disulfide bond formation in *Escherichia coli*. *Int. J. Biochem. Cell. Biol.* **38**, 1050–1062.

Misra, R., CastilloKeller, M., and Deng, M. (2000). Overexpression of protease-deficient DegP(S210A) rescues the lethal phenotype of *Escherichia coli* OmpF assembly mutants in a degP background. *J. Bacteriol.* **182**, 4882–4888.

Mogensen, J. E., and Otzen, D. E. (2005). Interactions between folding factors and bacterial outer membrane proteins. *Mol. Microbiol.* **57**, 326–346.

Nakada, S., Sakakura, M., Takahashi, H., Okuda, S., Tokuda, H., and Shimada, I. (2009). Structural investigation of the interaction between LolA and LolB using NMR. *J. Biol. Chem.* **284**(36), 24634–24643

Nakamoto, H., and Bardwell, J. C. (2004). Catalysis of disulfide bond formation and isomerization in the *Escherichia coli* periplasm. *Biochim. Biophys. Acta* **1694**, 111–119.

Nikaido, H. (2003). Molecular basis of bacterial outer membrane permeability revisited. *Microbiol. Mol. Biol. Rev.* **67**, 593–656.

Nuccio, S. P., and Baumler, A. J. (2007). Evolution of the chaperone/usher assembly pathway: Fimbrial classification goes Greek. *Microbiol. Mol. Biol. Rev.* **71**, 551–575.

Oguchi, Y., Takeda, K., Watanabe, S., Yokota, N., Miki, K., and Tokuda, H. (2008). Opening and closing of the hydrophobic cavity of LolA coupled to lipoprotein binding and release. *J. Biol. Chem.* **283**, 25414–25420.

Okon, M., Moraes, T. F., Lario, P. I., Creagh, A. L., Haynes, C. A., Strynadka, N. C., and McIntosh, L. P. (2008). Structural characterization of the type-III pilot-secretin complex from *Shigella flexneri*. *Structure* **16**, 1544–1554.

Olson, M. O., Nagabhushan, N., Dzwiniel, M., Smillie, L. B., and Whitaker, D. R. (1970). Primary structure of alpha-lytic protease: a bacterial homologue of the pancreatic serine proteases. *Nature* **228**, 438–442.

Pauwels, K., Lustig, A., Wyns, L., Tommassen, J., Savvides, S. N., and Van Gelder, P. (2006). Structure of a membrane-based steric chaperone in complex with its lipase substrate. *Nat. Struct. Mol. Biol.* **13**, 374–375.

Pauwels, K., Van Molle, I., Tommassen, J., and Van Gelder, P. (2007). Chaperoning Anfinsen: The steric foldases. *Mol. Microbiol.* **64**, 917–922.

Peters, R. J., Shiau, A. K., Sohl, J. L., Anderson, D. E., Tang, G., Silen, J. L., and Agard, D. A. (1998). Pro region C-terminus: Protease active site interactions are critical in catalyzing the folding of alpha-lytic protease. *Biochemistry* **37**, 12058–12067.

Qu, J., Behrens-Kneip, S., Holst, O., and Kleinschmidt, J. H. (2009). Binding regions of outer membrane protein A in complexes with the periplasmic chaperone Skp. A site-directed fluorescence study. *Biochemistry* **48**, 4926–4936.

Qu, J., Mayer, C., Behrens, S., Holst, O., and Kleinschmidt, J. H. (2007). The trimeric periplasmic chaperone Skp of *Escherichia coli* forms 1:1 complexes with outer membrane proteins via hydrophobic and electrostatic interactions. *J. Mol. Biol.* **374**, 91–105.

Ramm, K., and Pluckthun, A. (2000). The periplasmic *Escherichia coli* peptidylprolyl cis,trans-isomerase FkpA. II. Isomerase-independent chaperone activity in vitro. *J. Biol. Chem.* **275**, 17106–17113.

Ramm, K., and Pluckthun, A. (2001). High enzymatic activity and chaperone function are mechanistically related features of the dimeric *E. coli* peptidyl-prolyl-isomerase FkpA. *J. Mol. Biol.* **310**, 485–498.

Remaut, H., Rose, R. J., Hannan, T. J., Hultgren, S. J., Radford, S. E., Ashcroft, A. E., and Waksman, G. (2006). Donor-strand exchange in chaperone-assisted pilus assembly proceeds through a concerted beta strand displacement mechanism. *Mol. Cell.* **22**, 831–842.

Rhodius, V. A., Suh, W. C., Nonaka, G., West, J., and Gross, C. A. (2006). Conserved and variable functions of the sigmaE stress response in related genomes. *PLoS Biol.* **4**, e2.

Richarme, G., and Caldas, T. D. (1997). Chaperone properties of the bacterial periplasmic substrate-binding proteins. *J. Biol. Chem.* **272**, 15607–15612.

Rizzitello, A. E., Harper, J. R., and Silhavy, T. J. (2001). Genetic evidence for parallel pathways of chaperone activity in the periplasm of *Escherichia coli*. *J. Bacteriol.* **183**, 6794–6800.

Rosenau, F., Tommassen, J., and Jaeger, K. E. (2004). Lipase-specific foldases. *Chembiochem.* **5**, 152–161.

Rouviere, P. E., and Gross, C. A. (1996). SurA, a periplasmic protein with peptidyl-prolyl isomerase activity, participates in the assembly of outer membrane porins. *Genes Dev.* **10**, 3170–3182.

Sauer, F. G., Futterer, K., Pinkner, J. S., Dodson, K. W., Hultgren, S. J., and Waksman, G. (1999). Structural basis of chaperone function and pilus biogenesis. *Science* **285**, 1058–1061.

Sauer, F. G., Mulvey, M. A., Schilling, J. D., Martinez, J. J., and Hultgren, S. J. (2000). Bacterial pili: Molecular mechanisms of pathogenesis. *Curr. Opin. Microbiol.* **3**, 65–72.

Sauer, F. G., Pinkner, J. S., Waksman, G., and Hultgren, S. J. (2002). Chaperone priming of pilus subunits facilitates a topological transition that drives fiber formation. *Cell* **111**, 543–551.

Sauer, F. G., Remaut, H., Hultgren, S. J., and Waksman, G. (2004). Fiber assembly by the chaperone-usher pathway. *Biochim. Biophys. Acta* **1694**, 259–267.

Saul, F. A., Arie, J. P., Vulliez-le Normand, B., Kahn, R., Betton, J. M., and Bentley, G. A. (2004). Structural and functional studies of FkpA from *Escherichia coli*, a *cis/trans* peptidyl-prolyl isomerase with chaperone activity. *J. Mol. Biol.* **335**, 595–608.

Sauter, N. K., Mau, T., Rader, S. D., and Agard, D. A. (1998). Structure of alpha-lytic protease complexed with its pro region. *Nat. Struct. Biol.* **5**, 945–950.

Schafer, U., Beck, K., and Muller, M. (1999). Skp, a molecular chaperone of gram-negative bacteria, is required for the formation of soluble periplasmic intermediates of outer membrane proteins. *J. Biol. Chem.* **274**, 24567–24574.

Schlapschy, M., Grimm, S., and Skerra, A. (2006). A system for concomitant over-expression of four periplasmic folding catalysts to improve secretory protein production in *Escherichia coli*. *Protein. Eng. Des. Sel.* **19**, 385–390.

Shao, F., Bader, M. W., Jakob, U., and Bardwell, J. C. (2000). DsbG, a protein disulfide isomerase with chaperone activity. *J. Biol. Chem.* **275**, 13349–13352.

Shen, Q. T., Bai, X. C., Chang, L. F., Wu, Y., Wang, H. W., and Sui, S. F. (2009). Bowl-shaped oligomeric structures on membranes as DegP's new functional forms in protein quality control. *Proc. Natl. Acad. Sci. USA* **106**, 4858–4863.

Shibata, H., Kato, H., and Oda, J. (1998). Molecular properties and activity of amino-terminal truncated forms of lipase activator protein. *Biosci. Biotechnol. Biochem.* **62**, 354–357.

Shinde, U., and Inouye, M. (1993). Intramolecular chaperones and protein folding. *Trends Biochem. Sci.* **18**, 442–446.

Sklar, J. G., Wu, T., Kahne, D., and Silhavy, T. J. (2007). Defining the roles of the periplasmic chaperones SurA, Skp, and DegP in *Escherichia coli*. *Genes Dev.* **21**, 2473–2484.

Skorko-Glonek, J., Laskowska, E., Sobiecka-Szkatula, A., and Lipinska, B. (2007). Characterization of the chaperone-like activity of HtrA (DegP) protein from *Escherichia coli* under the conditions of heat shock. *Arch. Biochem. Biophys.* **464**, 80–89.

Slonim, L. N., Pinkner, J. S., Branden, C. I., and Hultgren, S. J. (1992). Interactive surface in the PapD chaperone cleft is conserved in pilus chaperone superfamily and essential in subunit recognition and assembly. *EMBO J.* **11**, 4747–4756.

Sohl, J. L., Jaswal, S. S., and Agard, D. A. (1998). Unfolded conformations of alpha-lytic protease are more stable than its native state. *Nature* **395**, 817–819.

Spiess, C., Beil, A., and Ehrmann, M. (1999). A temperature-dependent switch from chaperone to protease in a widely conserved heat shock protein. *Cell* **97**, 339–347.

Stirling, P. C., Bakhoum, S. F., Feigl, A. B., and Leroux, M. R. (2006). Convergent evolution of clamp-like binding sites in diverse chaperones. *Nat. Struct. Mol. Biol.* **13**, 865–870.

Stymest, K. H., and Klappa, P. (2008). The periplasmic peptidyl prolyl cis-trans isomerases PpiD and SurA have partially overlapping substrate specificities. *FEBS J.* **275**, 3470–3479.

Takeda, K., Miyatake, H., Yokota, N., Matsuyama, S., Tokuda, H., and Miki, K. (2003). Crystal structures of bacterial lipoprotein localization factors, LolA and LolB. *EMBO J.* **22**, 3199–3209.

Tapley, T. L., Korner, J. L., Barge, M. T., Hupfeld, J., Schauerte, J. A., Gafni, A., Jakob, U., and Bardwell, J. C. (2009). Structural plasticity of an acid-activated chaperone allows promiscuous substrate binding. *Proc. Natl. Acad. Sci. USA* **106**, 5557–5562.

Thanassi, D. G., Saulino, E. T., and Hultgren, S. J. (1998). The chaperone/usher pathway: A major terminal branch of the general secretory pathway. *Curr. Opin. Microbiol.* **1**, 223–231.

Thanassi, D. G., Stathopoulos, C., Karkal, A., and Li, H. (2005). Protein secretion in the absence of ATP: The autotransporter, two-partner secretion and chaperone/usher pathways of gram-negative bacteria (review). *Mol. Membr. Biol.* **22**, 63–72.

Tokuda, H. (2009). Biogenesis of outer membranes in gram-negative bacteria. *Biosci. Biotechnol. Biochem.* **73**, 465–473.

Tormo, A., Almiron, M., and Kolter, R. (1990). SurA, an *Escherichia coli* gene essential for survival in stationary phase. *J. Bacteriol.* **172**, 4339–4347.

Tran, A. X., Trent, M. S., and Whitfield, C. (2008). The LptA protein of *Escherichia coli* is a periplasmic lipid A-binding protein involved in the lipopolysaccharide export pathway. *J. Biol. Chem.* **283**, 20342–20349.

Truhlar, S. M., and Agard, D. A. (2005). The folding landscape of an alpha-lytic protease variant reveals the role of a conserved beta-hairpin in the development of kinetic stability. *Proteins* **61**, 105–114.

Ureta, A. R., Endres, R. G., Wingreen, N. S., and Silhavy, T. J. (2007). Kinetic analysis of the assembly of the outer membrane protein LamB in *Escherichia coli* mutants each lacking a secretion or targeting factor in a different cellular compartment. *J. Bacteriol.* **189**, 446–454.

Verger, D., Bullitt, E., Hultgren, S. J., and Waksman, G. (2007). Crystal structure of the P pilus rod subunit PapA. *PLoS Pathog.* **3**, e73.

Vetsch, M., Puorger, C., Spirig, T., Grauschopf, U., Weber-Ban, E. U., and Glockshuber, R. (2004). Pilus chaperones represent a new type of protein-folding catalyst. *Nature* **431**, 329–333.

Vetsch, M., Sebbel, P., and Glockshuber, R. (2002). Chaperone-independent folding of type 1 pilus domains. *J. Mol. Biol.* **322**, 827–840.

Walton, T. A., Sandoval, C. M., Fowler, C. A., Pardi, A., and Sousa, M. C. (2009). The cavity-chaperone Skp protects its substrate from aggregation but allows independent folding of substrate domains. *Proc. Natl. Acad. Sci. USA* **106**, 1772–1777.

Walton, T. A., and Sousa, M. C. (2004). Crystal structure of Skp, a prefoldin-like chaperone that protects soluble and membrane proteins from aggregation. *Mol. Cell.* **15**, 367–374.

Wandersman, C. (1989). Secretion, processing and activation of bacterial extracellular proteases. *Mol. Microbiol.* **3**, 1825–1831.

Watts, K. M., and Hunstad, D. A. (2008). Components of SurA required for outer membrane biogenesis in uropathogenic *Escherichia coli*. *PLoS One* **3**, e3359.

Webb, H. M., Ruddock, L. W., Marchant, R. J., Jonas, K., and Klappa, P. (2001). Interaction of the periplasmic peptidylprolyl *cis-trans* isomerase SurA with model peptides. The N-terminal region of SurA is essential and sufficient for peptide binding. *J. Biol. Chem.* **276**, 45622–45627.

Wodak, S. J., and Janin, J. (2002). Structural basis of macromolecular recognition. *Adv. Protein. Chem.* **61**, 9–73.

Wulfing, C., and Pluckthun, A. (1994). Protein folding in the periplasm of *Escherichia coli*. *Mol. Microbiol.* **12**, 685–692.

Xu, X., Wang, S., Hu, Y. X., and McKay, D. B. (2007). The periplasmic bacterial molecular chaperone SurA adapts its structure to bind peptides in different conformations to assert a sequence preference for aromatic residues. *J. Mol. Biol.* **373**, 367–381.

Yang, F., Gustafson, K. R., Boyd, M. R., and Wlodawer, A. (1998). Crystal structure of *Escherichia coli* HdeA. *Nat. Struct. Biol.* **5**, 763–764.

Zavialov, A., Zav'yalova, G., Korpela, T., and Zav'yalov, V. (2007). FGL chaperone-assembled fimbrial polyadhesins: anti-immune armament of gram-negative bacterial pathogens. *FEMS Microbiol. Rev.* **31**, 478–514.

Zavialov, A. V., Berglund, J., Pudney, A. F., Fooks, L. J., Ibrahim, T. M., MacIntyre, S., and Knight, S. D. (2003). Structure and biogenesis of the capsular F1 antigen from *Yersinia pestis*: Preserved folding energy drives fiber formation. *Cell* **113**, 587–596.

Zavialov, A. V., Tischenko, V. M., Fooks, L. J., Brandsdal, B. O., Aqvist, J., Zav'yalov, V. P., Macintyre, S., and Knight, S. D. (2005). Resolving the energy paradox of chaperone/usher-mediated fibre assembly. *Biochem. J.* **389**, 685–694.

Zheng, W. D., Quan, H., Song, J. L., Yang, S. L., and Wang, C. C. (1997). Does DsbA have chaperone-like activity? *Arch. Biochem. Biophys.* **337**, 326–331.

SEPARATE ROLES OF STRUCTURED AND UNSTRUCTURED REGIONS OF Y-FAMILY DNA POLYMERASES

By HARUO OHMORI,* TOMO HANAFUSA,* EIJI OHASHI,† AND CYRUS VAZIRI‡

*Institute for Virus Research, Kyoto University, 53 Shogoin-Kawaracho, Sakyo-ku, Kyoto 606-8507, Japan
†Department of Biology, Kyushu University, 6-10-1, Hakozaki, Higashi-ku, Fukuoka 812-8581, Japan
‡Department of Pathology, University of North Carolina, 614 Brinkhous-Bullitt Building, Chapel Hill, North Carolina 27599-7525

Abstract

All organisms have multiple DNA polymerases specialized for translesion DNA synthesis (TLS) on damaged DNA templates. Mammalian TLS DNA polymerases include Pol η, Pol ι, Pol κ, and Rev1 (all classified as "Y-family" members) and Pol ζ (a "B-family" member). Y-family DNA polymerases have highly structured catalytic domains; however, some of these proteins adopt different structures when bound to DNA (such as archaeal Dpo4 and human Pol κ), while others maintain similar

ADVANCES IN PROTEIN CHEMISTRY AND STRUCTURAL BIOLOGY, Vol. 78
DOI: 10.1016/S1876-1623(09)78004-0

99

structures independently of DNA binding (such as archaeal Dbh and *Saccharomyces cerevisiae* Pol η). DNA binding-induced structural conversions of TLS polymerases depend on flexible regions present within the catalytic domains. In contrast, noncatalytic regions of Y-family proteins, which contain multiple domains and motifs for interactions with other proteins, are predicted to be mostly unstructured, except for short regions corresponding to ubiquitin-binding domains. In this review we discuss how the organization of structured and unstructured regions in TLS polymerases is relevant to their regulation and function during lesion bypass.

I. Historical Background

For stable transmission of genetic information over generations, chromosomal and mitochondrial DNAs need to be replicated with extreme accuracy. Consistent with the requirement for replication accuracy, the three *Escherichia coli* DNA polymerases I, II, and III (discovered in 1954, 1970, and 1971, respectively) were found to exhibit high fidelity partly owing to the proofreading function of intrinsic 3′–5′ exonuclease activities (Kornberg and Baker, 1991). However, over very long time periods, genomes must adapt and acquire the changes that drive evolution. Thus, accumulation of point mutations in duplicated genes facilitates acquisition of new diversified functions for gene products ("No mutation, no evolution"). Under stressful conditions, it is particularly advantageous for mutation rates to increase, thereby giving rise to mutants likely to be better adapted to new environments (Radman, 1999).

Genetic studies in *E. coli* revealed that such an increase in mutation frequency requires active products of certain genes, so-called SOS-inducible genes such as *umuC* and *dinB* (alternatively named *dinP*) whose products show significant similarity to each other (for reviews, see Friedberg *et al.*, 2006). Based on the presumption that *E. coli* cells have only three DNA polymerases with proofreading function, the products of *umuC* (together with *umuD*) and *dinB* were assumed to interact with the replicative DNA polymerase III (Pol III) so as to decrease its fidelity when encountering a lesion on the template DNA. Similarly, genetic studies in *Saccharomyces cerevisiae* (sc) indicated that several genes (designated *REV* after *rev*ersion-less phenotype) are involved in both inducible and spontaneous mutagenesis (for a review, see Lawrence, 2004).

The yeast genes required for mutagenesis include *REV1, REV3*, and *REV7*. The *REV1* gene product shows similarity to the *E. coli* UmuC and DinB proteins, and the *REV3* gene product resembles the catalytic subunit of the replicative DNA polymerase δ. Furthermore, sequence analysis of *S. cerevisiae* genome revealed that the yeast has another gene homologous to the *E. coli umuC* and *dinB*, namely, *RAD30* (McDonald *et al.*, 1997; Roush *et al.*, 1998).

In 1996, Lawrence and his colleagues showed that the Rev3 and Rev7 proteins form an enzyme complex (designated Pol ζ for the sixth DNA polymerase found in *S. cerevisiae*) which performs replicative bypass of *cis–syn* T–T cyclobutane pyrimidine dimer (CPD), albeit inefficiently (Nelson *et al.*, 1996a). Additionally, these workers showed that the yeast Rev1 protein has an activity that inserts dCMP opposite an abasic site in template DNA (dCMP transferase activity) (Nelson *et al.*, 1996b). However, since the yeast Rev1 protein is much larger than the UmuC and DinB proteins [985 vs. 422 and 351 amino acid (aa) residues, respectively], it was unclear at the time whether or not the region of Rev1 resembling UmuC and DinB proteins corresponded to the dCMP transferase catalytic domain. In 1999, *in vitro* studies with purified DinB and UmuC (with or without UmuD', an active form of UmuD) proteins revealed that each of the proteins exhibits a DNA polymerase activity that is devoid of 3'–5' exonuclease activity. DinB and UmuC were designated Pol IV and Pol V, respectively (Reuven *et al.*, 1999; Tang *et al.*, 1999; Wagner *et al.*, 1999). Thus, the fourth and fifth *E. coli* DNA polymerases were identified over 25 years following discovery of the third replicative DNA polymerase III.

In the same year, the yeast Rad30 protein was found to bypass T–T CPD very efficiently and quite accurately by inserting two As opposite the dimer and the product was designated Pol η (the seventh DNA polymerase found in *S. cerevisiae*) (Johnson *et al.*, 1999a). More importantly in relationship to human diseases, the gene responsible to a cancer-prone syndrome, xeroderma pigmentosum variant (XP-V) was found to code for a human counterpart of Pol η (Johnson *et al.*, 1999b; Masutani *et al.*, 1999b). Furthermore, mammals have another homolog of the yeast Rad30 protein (designated Pol ι, Tissier *et al.*, 2000) as well as a homolog of the *E. coli* DinB protein (designated Pol κ, Ohashi *et al.*, 2000). Thus, together with a Rev1 homolog (Gibbs *et al.*, 2000), mammals have four similar DNA polymerases lacking proofreading function. These newly identified enzymes, all of which participate in translesion DNA synthesis (TLS), were classified as Y-family DNA polymerases in order to distinguish them from hitherto known families of DNA polymerases (A, B, C, and X) (Ohmori *et al.*, 2001).

Low-fidelity TLS polymerases represent a "double-edged sword." As evidenced by the fact that the XP-V patients lacking active Pol η are predisposed to skin cancer due to high incidence of mutations, the activity of human Pol η (hPol η) is required for decreasing mutations induced by UV irradiation. However, owing to intrinsic low fidelity, the action of Pol η or any other TLS polymerase inevitably confers an increase in the frequencies of mutations (McCulloch *et al.*, 2004). Nevertheless, higher organisms benefit from replication errors during "somatic hypermutations" (SHMs) that occur mainly at defined loci in the genes encoding immunoglobulin in immune cells. Indeed, Pol η has been shown to play a role in SHMs (for a review, see Weill and Reynaud, 2008). In addition to Y-family DNA polymerases, other error-prone DNA polymerases have been identified. For example, Pol μ and λ, which together with Pol β are classified as X-family polymerases, are involved in nonhomologous end joining in immune cells. Thus, error-prone DNA polymerases have important specialized roles *in vivo*.

In this review, we discuss structure-and-function relationship of Y-family DNA polymerases, focusing on motifs and domains that are important for interplay among Y-family DNA polymerases and their interactions with other proteins. The structures and functions of catalytic domains are more comprehensively discussed in other reviews (Prakash *et al.*, 2005; Yang and Woodgate, 2007). We now know that each TLS polymerase exhibits a characteristic pattern for bypassing different DNA lesions. A number of structural analyses of Y-family enzyme catalytic domains have deepened our understandings of their unique activities. All the structures of Y-family DNA polymerases have basic right-hand-like architecture consisted of "thumb," "palm," and "finger" domains with one additional domain termed "LF" (little finger), "PAD" (polymerase-associated domain) or "wrist". Y-family DNA polymerases have multiple motif sequences in common (Kulaeva *et al.*, 1996; Ohmori *et al.*, 1995). The five common motif sequences reside around the active sites in the tertiary structures near the incoming substrate site and the primer terminus.

However, the portions of the active sites, residing near the lesion-containing DNA template, are variable among the Y-family polymerases, consistent with the notion that each TLS polymerase copes with different species of DNA lesions.

Although a wealth of information is now available on the catalytic functions of TLS polymerases, very little is known about how these enzymes are recruited to sites of DNA damage *in vivo*. Our understanding of mechanisms of TLS polymerase recruitment was advanced considerably by the finding that the Rad6–Rad18 complex [an E2–E3 ubiquitin (Ub) ligase] modifies PCNA (proliferating cell nuclear antigen), the sliding clamp for DNA polymerases, via monoubiquitination in cells that acquire DNA damage (Hoege *et al.*, 2002; Stelter and Ulrich, 2003). Consistent with a role for monoubiquitinated PCNA (mUb-PCNA) in regulating TLS polymerase recruitment, each of the four Y-family polymerases has one or two copies of ubiquitin-binding domain (UBD) (Bienko *et al.*, 2005). Thus, direct interactions between mUb-PCNA and the UBD motifs may provide a mechanism for initial recognition of stalled replication forks by TLS polymerases. Additionally, Pol η, Pol ι, and Pol κ possess PCNA-interacting protein (PIP)-box sequences, further consistent with a PCNA-based mode of recruitment to sites of DNA damage. A PIP-box has not been identified in Rev1, although some workers have suggested that mouse Rev1 binds PCNA via its N-terminal BRCT domain (Guo *et al.*, 2006a). Thus it is possible that BRCT domain may serve as PIP-box substitutes in the context of Rev1. Another unique feature of Rev1 is that it interacts via its C-terminal domain (CTD) with Pol η, Pol ι, and Pol κ (Guo *et al.*, 2003; Ohashi *et al.*, 2004; Tissier *et al.*, 2004). The Rev1-interacting region (RIR) of Pol η, Pol ι and Pol κ is defined by short sequences in which the presence of two consecutive phenylalanine (F) residues is critical (Ohashi *et al.*, 2009). The Rev1 interaction is essential, at least for appropriate function of Pol κ. It is also known that Rev1-CTD interacts with the Rev7 accessory subunit of Pol ζ (Murakumo *et al.*, 2001), which is another TLS enzyme classified as a "B-family" DNA polymerase and believed to function mainly at the extension step after a DNA polymerase has inserted a base opposite DNA lesion.

Binding partners for Pol η have also been identified including Pol ι (Kannouche *et al.*, 2002), Msh2 (Wilson *et al.*, 2005), Rad18 (Watanabe *et al.*, 2004), Rad51 (McIlwraith *et al.*, 2005), and many other proteins

(Yuasa *et al.*, 2006), in addition to PCNA and Rev1-CTD. How are such interactions with many proteins accomplished and regulated? It seems likely that N-terminal halves of Pol η and other Y-family polymerases have tight tertiary structure suited for their respective catalytic functions, but the C-terminal halves involved for transient interactions with other proteins are mostly disordered. In general, intrinsically disordered proteins or inherently unfolded proteins are very common in eukaryotes, but less so in prokaryotes (see recent reviews, Dunker *et al.*, 2008; Fink, 2005). One advantage of disorder may be that multiple metastable conformations allow recognition of several targets with high specificity and low affinity. For any given protein it is more difficult to verify that a particular region is intrinsically disordered than to solve an ordered structure. Therefore, we necessarily rely on bioinformatics for predicting the presence of unstructured regions. Since information for protein folding is determined by primary aa sequence, information of nonfolding should also be specified by aa sequence. In fact, it is known that compared to sequences of ordered proteins, intrinsically disordered segments and proteins have significantly higher levels of certain aa's (E, K, R, G, Q, S, and P) and lower levels of others (I, L, V, W, F, Y, C, and N). Recently, a very useful method (DISOPRED2, http://bioinf.cs.ucl.ac.uk/disopred/) was developed for predicting disordered regions based on the primary sequence of proteins (Ward *et al.*, 2004). Employing this prediction program, we discuss structure-and-function relationships in Y-family polymerases.

This review is separated into two halves. In the first half, we describe structural aspects of each subgroup proteins among Y-family polymerases, trying to explain why they show different patterns of lesion bypass. Also, we focus on the differences in the structures of DNA-bound and DNA-unbound forms, pointing out flexibility in structures of TLS polymerases. In the second half, we discuss roles of protein–protein interactions, comparing binding motifs and domains present in Pol η, Pol ι, Pol κ, and Rev1 proteins. Among the five subfamilies in the Y-family DNA polymerases, the UmuC subfamily proteins are present only in bacteria and they are not discussed here. We first discuss the DinB/Pol κ subfamily proteins that are present in all three kingdoms of life, bacteria, archaea, and eukaryotes. Then we discuss the Pol η, Pol ι, and Rev1 subfamilies that are present only in eukaryotes.

II.   FLEXIBLE STRUCTURES OF Y-FAMILY DNA POLYMERASES

A.   DinB/Pol κ Subfamily, Enzymes Conserved in Bacteria, Archaea, and Eukaryotes

The DinB/Pol κ subgroup proteins are ubiquitously present from bacteria to humans, but notably absent in the completely sequenced genomes of *S. cerevisiae* and *Drosophila melanogaster*. The *E. coli* DinB protein was shown to have DNA polymerase activity (designated DNA polymerase IV or Pol IV), independently of accessory proteins such as UmuD′ and RecA which are required for UmuC-dependent DNA polymerase activity (designated Pol V). The structural aspects of Dpo4 (*DNA polymerase IV*) from the thermophilic archeon *Sulfolobus solfataricus* have been studied more extensively than *E. coli* Pol IV. To date, three different types of the Dpo4 structures have been reported, which are complexes with DNA and substrate (Ling *et al.*, 2001), apoenzyme (Wong *et al.*, 2008), and a complex with PCNA (Xing *et al.*, 2009). PCNA in archaea and eukaryotes or β-clamp in bacteria is a sliding clamp that encircles double-stranded (ds) DNA and tethers DNA polymerases to the primer terminus, thereby functioning as DNA polymerase processivity factor. Y-family polymerases are low-processivity enzymes, and their activities are probably restricted to limited regions near sites of DNA damage. Most (although may not all) Y-family polymerases possess motif(s) for binding to PCNA or β-clamp.

The crystal structures of Dpo4, first reported as a ternary complex with DNA and substrate, revealed that Dpo4 has some unique features, while having a right-hand-like shape consisting of finger, thumb, and palm domains, as observed for replicative DNA polymerases (Ling *et al.*, 2001; see Fig. 1A). The finger and thumb domains are significantly small, resulting in an open and spacious active site that accommodates various DNA lesions. In addition, the catalytic cores of Dpo4 and other Y-family polymerases have an additional domain called "LF" (Ling *et al.*, 2001), "PAD" (Trincao *et al.*, 2001), or "wrist" (Silvian *et al.*, 2001), which together with the other three catalytic core domains, constitutes a DNA-binding cleft. The open active site enables bypass of various DNA lesions, but together with the absence of proofreading exonuclease activity renders Y-family polymerases error-prone, especially when copying undamaged DNA templates. Subsequently, the Dpo4 apoenzyme structure was solved for a construct lacking the C-terminal 10 residues (342–351

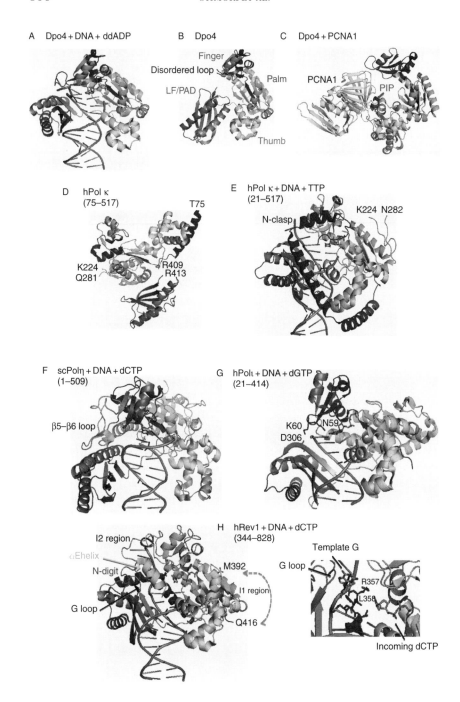

A   Dpo4 + DNA + ddADP

B   Dpo4

Finger
Disordered loop
Palm
LF/PAD
Thumb

C   Dpo4 + PCNA1

PCNA1
PIP

D   hPol κ
(75–517)
T75
K224
Q281
R409
R413

E   hPol κ + DNA + TTP
(21–517)
N-clasp
K224 N282

F   scPolη + DNA + dCTP
(1–509)
β5–β6 loop

G   hPolι + DNA + dGTP
(21–414)
K60 N59
D306

H   hRev1 + DNA + dCTP
(344–828)
I2 region
αEhelix
N-digit
M392
I1 region
G loop
Q416

Template G
G loop
R357
L358
Incoming dCTP

out of the total 352 residues) that were disordered in the above ternary complex. The apoenzyme structure revealed that Dpo4 undergoes a global conformational change with a large rotation (131°) of the LF domain relative to the three other domains upon DNA binding (Wong *et al.*, 2008, see Fig. 1B). Another difference between the apoenzyme and the ternary complex is that a loop (residues 34–39, see Fig. 1B) in the finger domain in the apo structure is disordered due to the absence of contacts made by the LF domain and DNA as observed in the DNA-bound structure.

At this point, it may be useful to compare the Dpo4 structures with those of Dbh (*DinB* homolog) from *Sulfolobus acidocaldarius* (previously described as *S. solfataricus* P1), which has 54% aa identity with Dpo4. The Dbh structures were first described as apo forms of the N-terminal catalytic fragment (2–216) (Zhou *et al.*, 2001) and the full-length protein (1–355) (Silvian *et al.*, 2001). In the apo structure of the full-length Dbh, the C-terminal domain ("wrist") projects from the end of the palm opposite the finger, differently from the extended form of the Dpo4 apoprotein. This is due to the interdomain contact mediated by the β9 strand of the wrist and the β5 strand of the palm (Silvian *et al.*, 2001). In the apo-Dbh structure, the C-terminal 10 residues and a loop region (35–39) in the finger domain are disordered, similarly as in the apo-Dpo4 structure (see Fig. 1B for the apo-Dpo4 structure). However, unlike Dpo4, the C-terminal wrist domain of Dbh does not undergo a large conformational change upon binding DNA, just rotating by 15–19° relative to the unliganded structure (Wilson and Pata, 2008). A large gap remaining between the finger and wrist domains provides ample space to accommodate an extra helical base in the template DNA chain, which permits misalignment of the template leading to deletion formation (Wilson and Pata, 2008). The *E. coli* DinB

FIG. 1.   Catalytic domain structures of Y-family polymerases. All molecular graphics images were prepared using MacPyMOL (http://www.pymol.org/), based on the PDB entries indicated in the parentheses. Y-family proteins are rainbow-colored; N-terminal regions are shown in blue and C-terminal regions in red. A, Dpo4 + DNA (1JX4); B, Apo–Dpo4 (2RDI); C, Dpo4 + PCNA1 (3FDS); D, Apo–hPol κ (1T94); E, hPol κ + DNA + TTP (2OH2); F, scPol η + DNA + dCTP (2R8J); G, hPol ι + DNA + dGTP (3GV8); H, hRev1 + DNA + dCTP (3GQC). In H, some portions of hRev1 around the template G and the incoming dCTP are enlarged. When truncated forms were used for structural analysis, the regions analyzed are denoted in the parentheses below each of the proteins. (See Color Insert.)

frequently makes single base deletions in runs of identical bases *in vivo* (Kim *et al.*, 1997) and Dbh shows a higher rate of single base deletion *in vitro* than Dpo4 (Potapova *et al.*, 2002; Boudsocq *et al.*, 2004). Thus, the position of the LF/wrist domain relative to the catalytic core domains appears to explain some functional differences observed between Dpo4 and Dbh.

While most eukaryotic and archaeal PCNA proteins are homotrimeric rings, three PCNA homologs (PCNA1, PCNA2, and PCNA3) exist in *S. solfataricus* and the functional form of PCNA in the archaeon is a hetero-trimer consisted of the three homologs. Interestingly, Dpo4 interacts with PCNA1 alone, mainly through the C-terminal sequence 342-EAIGLDKFFDT-352 (the conserved residues are underlined) that is similar to the consensus sequence of PIP-box Qxx(M,L,I)xxFF (x is any residue) (Warbrick, 1998). Structural analysis of Dpo4 in complex with the PCNA1–PCNA2 heterodimer (Dpo4–p12) revealed that the PCNA-bound Dpo4 is in an extended conformation, in which the LF domain is rotated from the other three domains, differently from the apoprotein (Xing *et al.*, 2009, see Fig. 1C). In the Dpo4–p12 complex, the LF is at the top of the front ring surface, being dissociated from the thumb, palm, and finger domains that are located at the side surface of PCNA ring. Some residues in the finger, thumb, and LF domains are involved in conformation-dependent interactions with PCNA1, in addition to the main interaction by the C-terminal PIP-box. In the PCNA-bound struc-ture, the DNA-binding cleft is disrupted because the LF stays near the central cavity of the PCNA ring structure, but the other three domains are kept away from it.

The above results predict at least two major conformational changes in the Dpo4 structure upon binding to PCNA and DNA, respectively. Two flexible hinge regions (hinge 1 and 2) were identified in Dpo4. The hinge 1 (residues 234–243) is at the linker located between LF and the other three catalytic core domains (thumb, palm, and finger), which affects the position and orientation of LF relative to the core domains. The hinge 2 exists at the C-terminus of LF and confers different orientations of LF and other domains relative to the PCNA ring. Such flexibility appears to be critical for Dpo4 to adopt different conformations even when bound to PCNA; a "carrier configuration" not competing with the replicative DNA polymerase for binding to the primer terminus and an "active configura-tion" engaged for DNA synthesis when the replicative enzyme is stalled at the site of DNA damage (Xing *et al.*, 2009).

Eukaryotic Pol κ homologs are much larger than Dpo4 that is essentially consisted of catalytic domains, with extensions at both N- and C-termini that confer additional domains for interaction with DNA and other proteins. The N-terminal extension of human (h) Pol κ (~100 residues in length, called the "N-clasp") encircles DNA together with the palm, finger, thumb, and LF/PAD domains, thus explaining why the N-terminal portion is important for binding to DNA (Lone *et al.*, 2007, see Fig. 1E). The structure of the apo protein was determined with a form lacking the N-clasp (hPol κ_{69-526}), in which the PAD tucks under and behind the palm domain, not forming a DNA-binding cleft with the other three domains (Uljion *et al.*, 2004, see Fig. 1D). In a ternary complex of a longer form (hPol κ_{19-526}) with DNA and substrate, the PAD docks directly in the major groove of DNA, moving ~50 Å from the position in the apoenzyme (Lone *et al.*, 2007). This suggests that the linker region between LF/PAD and the other catalytic core domains is flexible in hPol κ, as described above for Dpo4. Thus, we conclude that Dpo4 and hPol κ are "convertible" between two different forms, DNA-bond and DNA-unbound ones.

Whether Pol κ functions solely as an "extender" to extend from the nucleotide inserted opposite a lesion by another polymerase (Prakash *et al.*, 2005) or also as an "inserter" for certain species of DNA lesion (Ohmori *et al.*, 2004) is a controversial issue. Several groups have shown that hPol κ is capable of inserting the correct dCMP opposite N^2-benzo[a] pyrene diol epoxide (BPDE)–adducted dG (N^2-BPDE–dG) adducts, while very inefficiently inserting a base opposite N^6-BPDE–dA adducts (Zhang *et al.*, 2000; Rechkoblit *et al.*, 2002, Suzuki *et al.*, 2002). Choi *et al.* (2006) showed that hPol κ could bypass bulky N^2-alkyl–dG adducts of increasing size accurately, but the efficiencies of dCMP insertion opposite such N^2-dG lesions were still lower when compared with that opposite nondamaged dG template. In contrast, Jarosz *et al.* (2006) reported that mouse Pol κ bypassed N^2-furfuryl–dG adduct twofold more efficiently than nondamaged dG. Determining the structure of a ternary complex of hPol κ_{19-526}, DNA (not containing any lesion) and a substrate (dTTP), Lone *et al.* (2007) found that hPol κ has a constricted active site, not large enough to accommodate two damaged bases simultaneously. These workers proposed that the "extender" activity of hPol κ could be explained by the constricted active site and DNA encirclement by N-clasp. In contrast, from modeling studies based on the same structure,

Jia *et al.* (2008) argued that the N-clasp of hPol κ could favor base paring of dCTP with N^2-BPDE–dG adduct and disfavor any dNTP incorporation opposite N^6-BPDE–dA adduct.

The three-dimensional structure of hPol $κ_{19-526}$ is very similar to that of Dpo4, and the two proteins are mostly superimposable, the biggest difference being the presence of N-clasp in hPol $κ_{19-526}$. Nevertheless, Dpo4 and hPol $κ_{19-526}$ show clear difference in the bypass of 7,8-dihydro-8-oxoguanine (8-oxoG); Dpo4 incorporates dCTP more efficiently than dATP opposite 8-oxoG (Rechkoblit *et al.*, 2006; Zang *et al.*, 2006), but hPol κ incorporates dATP more efficiently than dCTP (Haracska *et al.*, 2002; Irimia *et al.*, 2009; Jaloszynski *et al.*, 2005; Zhang *et al.*, 2000). Two groups recently succeeded in solving the structures of hPol $κ_{19-526}$ inserting dATP opposite 8-oxoG (Carpio *et al.*, 2009; Irimia *et al.*, 2009) and found that the *syn* conformation of 8-oxoG for Hoogsteen base paring with incoming dATP is stabilized by the Met135 residue in the finger domain of hPol κ. Within the active site of Dpo4, the *anti* conformation of 8-oxoG for base pairing with dCTP is stabilized mainly due to the presence of the Arg322 residue in the LF/PAD domain, but the corresponding Leu508 residue in the LF/PAD of hPol κ allows 8-oxoG to keep the *syn* conformation. Thus, error-free or error-prone bypass of a DNA lesion by TLS polymerases appears to be determined by subtle differences in the structures surrounding the substrate and the damaged template.

The disorder profiles of Dpo4 and hPol κ obtained by using the DISOPRED2 program are shown in Fig. 2A and B, respectively. Figure 2A suggests that the entire region of Dpo4 has very low disorder probabilities, except for the C-terminal PIP-box region. As described above, the structure of DNA-bound Dpo4 indicated that only a small portion at the C-terminus is disordered. Therefore, the structure of apo-Dpo4 molecule was determined by crystallizing a truncated form that lacks the 10 C-terminal residues (342–351). Subsequent analysis of the Dpo4–p12 complex structure showed that upon binding to PCNA1, the C-terminal PIP-box is well structured, except for the very last residue Thr352. It should be noted that a small peak around 240 in Fig. 2A corresponds to the hinge 1 in Dpo4 (residues 234–243). Figure 2B suggests that the catalytic domain (100–520) of hPol κ has low disorder probabilities throughout the entire region, similarly to Dpo4. Here again, a small peak around aa 410 corresponds to the flexible linker region between LF/PAD and the three other catalytic core domains. Thus far, several different

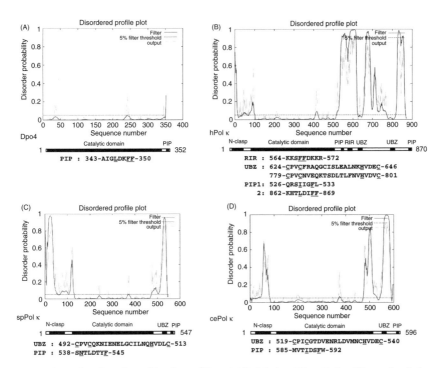

FIG. 2.   Disordered profile plots of Dpo4 (A), hPol κ (B), spPol κ (C), and cePol κ (D). Each of the plots was obtained from the DISOPRED2 server (http://bioinf.ucl.ac. uk/diospred/), after inputting the entire primary sequence of the respective protein. The aa sequences of domains and motifs are shown below the plot, in which conserved residues are underlined.

forms of hPol κ (apoprotein or complexes with damaged or undamaged DNA) have been structurally analyzed, all of which indicate that an internal region (225–281) is unstructured. However, the disorder profile shown in Fig. 2B does not predict the presence of such an unstructured region, implying that the prediction by the DISOPRED2 program cannot detect every unstructured region. Because hPol κ$_{69-526}$ was the only N-clasp-deficient construct successfully crystallized (Lone *et al.*, 2007), the N-clasp region was suggested to be disordered in the absence of DNA. Figure 2B shows that the N-terminal region up to 100 has higher disorder probabilities, with a peak around 90 probably corresponding to the linker region between the N-clasp and the thumb domain.

In contrast to the rigid catalytic domain, the C-terminal half of hPol κ is predicted to have multiple regions of high or low disorder probabilities. The two "valley" regions around amino acids 630 and 790 with very low disorder probabilities correspond to the ubiquitin-binding zinc-finger (UBZ) domains. The consensus sequence of RIR is denoted by xxxFFyyyy (x, no specific residue and y, no specific residue but not proline) (Ohashi *et al.*, 2009). The hPol κ sequence has two sites containing FF, one at 567–568 in RIR and the other at 868–869 in the C-terminal PIP-box sequence. In contrast to the two UBZs, the RIR site around FF567–568 appears to be embedded in one of unstructured regions. Although the disordered profile plot suggests that the extreme C-terminal region of hPol κ is "ordered," it is likely that the C-terminal PIP-box assumes the characteristic 3_{10} helix structure upon binding to PCNA (see Hishiki *et al.*, 2009), as described above for the C-terminal PIP-box sequence in Dpo4. It is noted that in the DISOPRED2 prediction, false assignment of order can occur as a result of stabilizing interactions by other macromolecules. Very recently, another PIP-box-like sequence (526-QRSIIGFL-533) was found to have a PCNA-binding activity by yeast two-hybrid assay (our unpublished observation). Since this sequence is located immediately downstream of the catalytic domain (a similar arrangement to the sole PIP-box of hPol ι and the PIP1 site of hPol η as described below), the internal PIP-box of hPol κ is designated PIP1 and the C-terminal PIP-box PIP2.

While the disorder profile of the *E. coli* DinB is very similar to that of Dpo4, those of the *Schizosaccharomyces pombe* and *Caenorhabditis elegans* Pol κ homologs (spPol κ and cePol κ, 547 and 596 residues in the total, respectively) are of intermediate disorder when compared with prokaryotic and human DinB/Pol κ homologs, as shown in Fig. 2C and 2D. Both spPol κ and cePol κ have N-terminal expansions similar to hPol κ and C-terminal extensions much shorter than that of hPol κ. Both spPol κ and cePol κ have a single copy of UBZ, as well as a PIP-box sequence at the C-terminus.

B. Pol η Subfamily, Versatile Enzymes for Correct Bypass of a Variety of DNA Lesions

Pol η homologs are found from *S. cerevisiae* (encoded by the *RAD30* gene) to humans (encoded by the *XPV* gene), almost in every eukaryotic organism whose genome sequence has been determined, but not in prokaryotes. The *E. coli* Pol V (encoded by the *umuC* gene) and its homologs are considered to be functional counterparts in bacteria, because Pol V and Pol η are able to cope with a wide variety of DNA lesions. Pol η is unique because it is able to replicate through T–T CPD, by inserting two As opposite the lesion at the same efficiency and accuracy as that with undamaged template DNA. Furthermore, it is able to bypass a variety of DNA lesions by inserting correct base(s) opposite them at different efficiencies depending on species of DNA lesions (Masutani *et al.*, 2000). Nevertheless, Pol η is not always error-free and it sometimes commits error-prone bypass of certain DNA lesions (for more detailed descriptions of Pol η-mediated error-prone bypass, see Vaisman *et al.*, 2004).

The most characteristic feature of Pol η is its ability to replicate through UV-induced CPDs efficiently and accurately (Johnson *et al.*, 1999a,b; Masutani *et al.*, 1999a), suggesting that the active site should be large enough to accommodate such cross-links. However, Pol η is unable to bypass through another major UV-induced adduct (6–4) photoproducts [(6–4)PP], while it incorporates one of the four dXTPs opposite the 3'-T of T–T (6–4)PP, not extending further (Masutani *et al.*, 1999a). In general, TLS is efficient for those DNA lesions (such as CPDs) which have small distortion in the DNA structure still allowing Watson–Crick base paring, but it is inefficient for those [such as (6–4)PP] making large distortion, most of which should be rapidly removed by nucleotide excision repair (Masutani *et al.*, 1999b).

The structure of hPol η, as either an apoprotein or a complex with DNA, has not been reported. Therefore we discuss structure–function aspects of hPol η based on the reported structures of *S. cerevisiae* Pol η (scPol η, see Fig. 1F). The catalytic core structure of scPol η apoprotein was first determined with a truncated form lacking the C-terminal 119 aa residues that contained the UBD (termed UBZ) and the PIP-box (Trincao *et al.*, 2001). Subsequently, ternary complexes of scPol η with DNA containing cisplatin-induced 1,2-d(GG) adducts and dCTP were solved (Alt *et al.*, 2007). Comparison of the ternary complexes with the apoprotein revealed small motions of the thumb and PAD domains toward DNA resulting in more intimate interaction with DNA (Alt *et al.*, 2007). This implies that the catalytic core of scPol η apoprotein has a tight structure, in contrast to Dpo4 and hPol κ that undergo large conformational changes upon DNA binding. While scPol η also has a relatively long linker region (15 aa residues in length) between the thumb and LF/PAD domains, it is not as flexible as the hinge 1 (10 aa residues) in Dpo4, probably due to the close contacts between the finger and PAD domains in the apoprotein structure. Thus, we may conclude that scPol η has a tightly "preassembled" structure regardless of DNA binding, and does not interconvert between "carrier" and "active" conformations as proposed above for Dpo4. However, it is unknown whether hPol η is "preassembled" or "convertible" type. In scPol η, the long loop region between the β5 and β6 sheets in the finger domain (see Fig. 1F) is apparently involved in the contact between the finger and PAD domains, but the sequence corresponding to the loop is much shorter in hPol η (Trincao *et al.*, 2001).

It is generally believed that high-fidelity replicative DNA polymerases make errors much less frequently than TLS polymerases devoid of proof-reading function. There is one exceptional case for bypass of 8-oxoG. Replicative DNA polymerases replicate through 8-oxoG by inserting dAMP more frequently than dCMP (Shibutani *et al.*, 1991), but some TLS polymerases, for example Dpo4 as described above, insert dCMP in preference to dAMP opposite 8-oxoG. The enzyme scPol η exhibits 20-fold higher efficiency for incorporation of dCMP over dAMP opposite 8-oxoG, which contributes to preventing spontaneous G:C to T:A trans-versions caused by 8-oxoG in *ogg1*-defective mutants of *S. cerevisiae* (Haracska *et al.*, 2000). In the absence of the Ogg1 glycosylase to remove 8-oxoG paired with C, replicative polymerases have increased chances to encounter 8-oxoG and replicate past it by frequently inserting A opposite the lesion. The resulting 8-oxoG/A is recognized by the mismatch repair system, which removes a DNA fragment containing A in pairing with 8-oxoG. If the subsequent DNA synthesis is mediated by scPol η, error-free bypass across 8-oxoG contributes to the suppression of G:C to T:A mutations. Interestingly, similar to other modes of TLS, this process also depends on monoubiquitination of PCNA by the Rad6–Rad18 complex and on the UBZ and PIP-box functions in scPol η (de Padula *et al.*, 2004; Auffret van der Kemp *et al.*, 2009). If unmodified PCNA recruits Pol δ to fill a gap containing 8-oxoG, dAMP may be incorporated again opposite the lesion. Thus, error-free or error-prone DNA synthesis is not simply determined by the presence of a DNA lesion, rather, it depends on the combination of a lesion and DNA polymerase involved in its bypass.

As shown in Fig. 3A and B, the N-terminal catalytic domains of scPol η and hPol η have low disorder probabilities, whereas the C-terminal regions of both are predicted to be mostly disordered, except for the regions

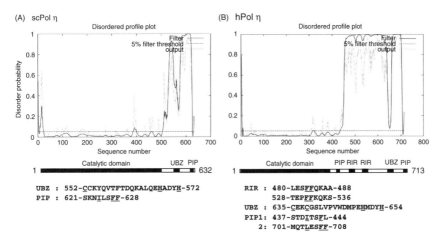

(A) scPol η

UBZ : 552-CCKYQVTFTDQKALQEHADYH-572
PIP : 621-SKNILSFF-628

(B) hPol η

RIR : 480-LESFFQKAA-488
       528-TEPFFKQKS-536
UBZ : 635-CEKCGSLVPVWDMPEHMDYH-654
PIP1: 437-STDITSFL-444
    2: 701-MQTLESFF-708

FIG. 3.   Disordered profile plots of scPol η (A) and hPol η (B).

corresponding to the single UBZ domain and the PIP-box at the C-terminus. Acharya *et al.* (2008) suggested that hPol η might have a PIP-box (437-STDITSFL-444, designated PIP1) immediately downstream of the catalytic domain, which is distantly related to the PIP-box consensus sequence Qxx(I,L,M)xxFF. However, we could not see any signal for PCNA binding by PIP1 when the 430–449 sequence of hPol η was examined for interaction with PCNA by yeast two-hybrid assay (Fig. 4A). As expected, we detected strong signals for PCNA binding with the C-terminal PIP-box sequence of hPol η (designated PIP2) and the PIP-box sequence of hPol ι under the same conditions. Furthermore, we could not see any significant increase on the PCNA interaction even when we introduced L444F substitution to make the PIP1 more similar to the consensus sequence of PIP-box (data not shown). Nevertheless, we cannot exclude a possibility that the PIP1 of hPol η may require some additional sequence to exhibit its PCNA-binding activity. The region near the C-terminus corresponding to the PIP2 sequence is predicted to be "ordered"; however, the region is expected to be "disordered" without binding to PCNA, as discussed above for the C-terminal PIP-box sequences of Dpo4 and hPol κ.

While hPol κ (and hPol ι, as described below) has a single RIR, hPol η has two RIRs around FF483–484 and FF531–532 (Ohashi *et al.*, 2009), both of which are embedded in a large unstructured region (see Fig. 3B). Each of the two RIRs independently binds to

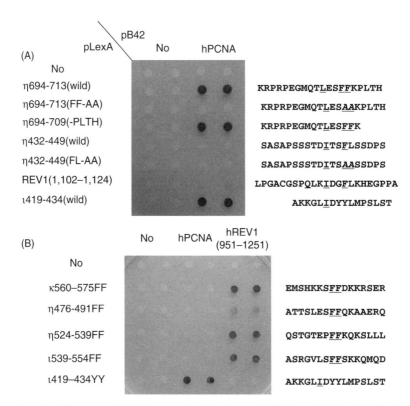

FIG. 4. Yeast two-hybrid assay for binding to PCNA and hRev1-CTD. Each segment of hPol η, hPol ι, hPol κ, or hRev1 was inserted into pLexA (BD) vector and the entire region of PCNA or a region (951–1,251) of hRev1 was inserted into pB42AD vector. The aa sequences of the inserted segments are shown in the right, in which conserved residues are underlined and altered sequences are shown in Italic. "No" indicates empty vector. Experiments were performed as described previously (Ohashi *et al.*, 2009).

hRev1-CTD. In the case of hPol κ, FF567–568AA substitution in the RIR sequence completely abolished its hRev1-binding activity and the FF567–568AA mutant could not correct the BPDE sensitivity of *Polk*-defective mouse embryonic fibroblast (MEF) cells (Ohashi *et al.*, 2009). In contrast, the mutant of hPol η carrying both FF483–484AA and FF531–532AA substitution that lost most of the hRev1-binding activity could fully complement the UV sensitivity of XP-V cells, as the wild type did (Akagi *et al.*, 2009). The double mutant also suppressed

UV-induced mutations in XP-V cells, but it reduced spontaneous mutations only partially. These results imply that the interaction with hRev1 is dispensable for accurate bypass of UV-induced lesions by hPol η, yet it might be necessary for bypass of some DNA lesions, which probably depend on coordinated actions by multiple TLS polymerases (see below for further discussions).

Another interesting feature of hPol η is that it interacts with many other proteins. Watanabe *et al.* (2004) showed that Rad18 directly interacts with hPol η and helps it form nuclear foci in UV-irradiated cells. The C-terminal 158 aa region of hPol η (from 556 to the C-terminus) containing the UBZ domain and the PIP2 site was sufficient for the Rad18 interaction. Furthermore, Kannouche *et al.* (2002) reported that hPol ι interacted mainly with the 352–595 region (and also weakly with the 595–713 region) of hPol η and concluded that hPol η is required for targeting hPol ι to the replication machinery. In these two cases, the interaction sites are not precisely mapped and no mutants specifically affecting such interactions are available. The significance of these interactions will be discussed in a later section.

C. Pol ι Subfamily, Enigmatic Enzymes Inserting G Opposite Undamaged T

Pol ι subfamily proteins are present only in eukaryotes including mammals, but lacking in *S. cerevisiae* or *S. pombe* while present in *Drosophila melanogoster* and *Neurospora crossa* (for recent progresses on the *in vivo* functions of Pol ι, see a review by Vidal and Woodgate, 2009). Human and mouse Pol ι have been thought to contain 715 and 717 residues, respectively; however, another in-frame ATG codon was recently found to exist in the upstream of the presumed start codon in each case. Furthermore, because the extended sequences of Pol ι homologs in mammals are well conserved (Fig. 5), it seems more likely that human and mouse Pol ι proteins are longer than originally thought. Interestingly, there is heterogeneity within the extended coding region in human genomic and cDNA sequences, because of the presence of CGA repeat. The sequence with four CGA repeats has an N-terminal extension of 25 amino acids and that with three CGA repeats has the extension of 24 amino acids which is completely identical with that of chimpanzee.

```
Human          MEKLGVEPEEEGGGDDDEEDAEAWAMELADVGAAASSQG
Chimpanzee     MEKLGVEPEEEGGGDD-EEDAEAWAMELADVGAAASSQG
Bovine         MEKRGVEPEEAGGGD--EEADETRAMERAEAGAPGGWPG
Rat            ---MGVEPEEEGGPA--EEEDFSRAMEPSDSGAPGGSRA
Mouse          ---MGVESEEEGGPA--EEEDAPRAMEPLHAGAAGSSRA
```

```
        Met Glu Lys Leu Gly Val Glu Pro Glu Glu Glu Gly Gly
   740  atg gag aag ctg ggg gtg gag ccg gag gag gaa ggc ggc

   739  atg gag aag ctg ggg gtg gag ccg gag gag gaa ggc ggc


        Gly Asp Asp Asp Glu Glu Asp Ala Glu Ala Trp Ala Met
   740  ggc gac gac gac gag gaa gac gcc gag gcc tgg gcc atg

   739  ggc gac gac --- gag gaa gac gcc gag gcc tgg gcc atg
```

FIG. 5.   Multiple alignment of the N-terminal regions in mammalian Pol ι homologs. In the upper aa sequence alignments, the Met residues of human and mouse Pol ι that were previously assigned as the first residue are underlined. In the lower sequences, the regions from the newly assigned Met start codon to the previously assigned one in the gene coding for the 740 or 739 aa protein are shown for human Pol ι, and are derived from the NCBI entries NM_007195 and AK301578, respectively. Many cDNA clones with the 5′-end sequence identical to either one of the two entries are found in human EST libraries at almost equal frequencies, implying that the difference is due to heterogeneity, not to sequence error. CGA repeats in the newly identified N-terminal sequences are denoted by a horizontal line with an arrow.

A unique feature of hPol ι is that the enzyme shows extremely high error rates when replicating on template pyrimidines, while showing accurate replication on template purines. For example, hPol ι misinserts G opposite T 3–10 times more frequently than the correct A. Structures of hPol ι have been studied, all as binary complexes with DNA or ternary complexes with DNA and substrate, but none as apoenzyme. Nair *et al.* (2004, 2005a) reported that in a ternary complex of hPol ι with DNA and an incoming substrate, the template purines (A or G) adopt a *syn* conformation for Hoogsteen base pairing. It explained the mechanism by which hPol ι could bypass $1,N^6$-ethenoadenine and N^2-ethylguanine by inserting the correct T and C, respectively, because both of the adducts cannot form the normal Watson–Crick base paring if they keep *anti* conformation (Nair *et al.*, 2006a; Pence *et al.*, 2009). While the template purines form *anti* conformation in the binary complex of hPol ι and DNA, the incoming substrate imposes an *anti* to *syn* conformational change because hPol ι has a constricted active site (Nair *et al.*, 2006b). In such a constricted active site, Watson–Crick base paring that requires the $C1'–C1'$ distance of 10.5 Å is disfavored and Hoogsteen base paring that requires the distance of around 8.5 Å is favored. When T is the template base, it remains in an *anti* conformation and the selection of dGTP over dATP is partly due to the hydrogen bonding between the N^2 amino of dGTP and Gln59 (the numbering of residues is followed as the old ones for the total length of 715 residues) in the finger domain of hPol ι (Kirouac and Ling, 2009; Jain *et al.*, 2009).

The results by Kirouac and Ling (2009) explain well why hPol ι has a narrowed active site (Fig. 1G). The loop between β2 and β3 in the finger domain of hPol ι is much shorter, for example, when compared with that of Dpo4. The residue Lys60 in the top of the loop interacts with Asp306 in the LF/PAD and Tyr61, together with Ser307 and Arg347 in the LF/PAD, interacts with the template DNA strand, thereby constituting a lid to the template DNA chain. The other end of the active site is defined by the three invariant acidic residues (Asp34, Asp126, and Glu127), which are essential for phosphodiester bond formation. Thus, the narrowed active site limits the $C1'–C1'$ distance of the replicating base pair to within 9 Å in hPol ι. Furthermore, hPol ι has amino acid residues with relatively large side chains in the finger domain that contact the replicating base pair in the active site. Among them, Gln59, which is conserved among Pol ι subgroup proteins, forms a unique hydrogen bond with the incoming dGTP that maintains an *anti* form, thereby facilitating the misincorporation of dGTP opposite template T.

Since no apo form of hPol ι has been reported, we may only speculate whether hPol ι is a "convertible" or "preassembled" type, based on the structures of the binary complex with DNA. As described above, the interaction between the finger and LF/PAD domains of hPol ι appears to be very weak, mostly depending on the interaction between Lys60 in the finger domain and Asp306 in the LF/PAD. The interaction is probably facilitated by binding of Ser307 and Arg347 in the LF/PAD to template DNA chain. Since there seems no close contact between the finger domain and LF/PAD (as seen in the case of scPol η, Fig. 1F), it is likely that apo form of hPol ι may have a conformation different from the DNA-bound form.

It has been thought that hPol ι does not have an N-terminal extension ahead of the catalytic domain; however, we now know that hPol ι has an N-terminal extension, which has relatively high disorder probabilities peaking around residue 40 (corresponding to the new numbering of full-length 740

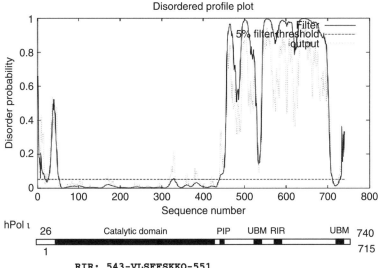

FIG. 6. Disordered profile plot of hPol ι. The entire 740 aa hPol ι sequence including the newly identified N-terminal 25 aa's was analyzed for disorder probability, but the positions of the motifs and domains are presented using the old numbering system (reflecting the N-terminally truncated 715 aa species). The old numbering may be converted to the new full-length sequence by addition of 25.

residues), as shown in Fig. 6. Thus far, structures of hPol ι have been analyzed with a truncated species comprising residues 26–445 (corresponding to 1–420 of the old numbering system). Such analyses indicate that N-terminal residues 26–50 of hPol ι (1–25 of the old numbering system) are disordered. The N-terminal extension of hPol ι is shorter than the length of the N-clasp in hPol κ (70–100 residues) and very similar in length to the N-digit of hRev1 (around 40–50 residues). Therefore, hPol ι may have an N-terminal domain, similar to the N-digit of hRev1.

The PIP-box sequence of hPol ι exists immediately downstream of the catalytic domain. Until recently, the PIP-box sequence of hPol ι has been thought to be 420-KKG<u>L</u>IDYY-427, in which YY supposedly corresponds to FF in the consensus sequence of PIP-box (Haracska *et al.*, 2005; Vidal *et al.*, 2004). However, more detailed *in vitro* PCNA-binding assays using peptides of altered sequences and structural determination of peptide-bound PCNA revealed that 421-KGL<u>I</u>DYYL-428 corresponds to the PIP-box consensus sequence (Hishiki *et al.*, 2009). The region downstream of the catalytic domain is predicted to be mostly unstructured, except for the two ubiquitin-binding motifs (UBMs). UBZs in hPol η or hPol κ were originally recognized to be zinc-finger motifs and were subsequently found to bind Ub. However, UBMs in hPol ι and Rev1 were originally identified as UBDs in screens for proteins that interact with Ub and a mutant form of Ub in which Ile44 (critical for binding to many proteins) was substituted to Ala (Bienko *et al.*, 2005). UBMs are predicted to comprise two α-helices separated by the central proline residue conserved among them. The entire sequence of hPol ι has only one FF site at 547–548 and therefore the flanking sequence (540-SRGVLSFF-548) was once presumed to be a PCNA-binding site (Haracska *et al.*, 2001). However, FF547–548AA substitution of hPol ι did not affect PCNA-binding (Vidal *et al.*, 2004), yet completely abolished the hRev1-CTD binding (Ohashi *et al.*, 2009) demonstrating that FF547–548 is an RIR. Here again, the RIR in hPol ι is located in a region of high disorder probabilities, similar to RIRs in hPol κ and hPol η. The C-terminal 224 residues of hPol ι were shown to be required for the interaction with hPol η (Kannouche *et al.*, 2002), but the precise interaction site has not been identified yet.

D.   *Rev1 Subfamily, dCMP Transferase with Noncatalytic Function Important for In Vivo TLS*

Rev1 proteins, together with Pol ζ (a complex consisting of the Rev3 catalytic subunit and the Rev7 accessory subunit) and Pol η, have been found in most eukaryotic organisms whose genomes have been sequenced. Rev1 proteins are composed of three portions, the N-terminal portion containing a BRCT domain, the central catalytic domain, and the C-terminal portion containing multiple motifs for interactions with other proteins. Yeast Rev1 was the first member of all the Y-family proteins that was found to have a catalytic activity, but it utilized only dCTP as substrate to insert opposite abasic sites, dU and undamaged dG (Nelson *et al.*, 1996b). The inserted dC was extended efficiently by the yeast Pol ζ. Furthermore, a mutant (*rev1-1* carrying the G193R substitution in the N-terminal BRCT domain) that retained the dCMP transferase activity was found to be defective for bypass of T–T (6–4)PP but proficient for bypass of T–T CPD (Nelson *et al.*, 2000). Thus, Rev1 was thought to have a "second" role other than dCMP transferase, probably for coordinated action with Pol ζ that is essentially required for bypass of abasic and (6–4)PP lesions.

While all Y-family DNA polymerases have five common motifs within the catalytic domains, Rev1 subfamily proteins share one additional motif containing the SRLHH sequence at the N-ternmius of the catalytic domain. The additional motif was necessary for the human Rev1 to exhibit dCMP transferase activity *in vitro* (Masuda *et al.*, 2001). Structural analysis of the yeast Rev1 catalytic core in complex with DNA and incoming dCTP revealed that the template G is evicted from the DNA helix by the Leu residue in the additional motif and the adjacent residue Arg interacts with the incoming dCTP (Nair *et al.*, 2005b). The additional sequence of Rev1 catalytic domain constitutes an extra domain called "N-digit" in the tertiary structure, which occupies a space between the palm and PAD. The Rev1 PAD has a relatively long loop, designated "G loop," which accommodates evicted template G in preference to other template bases. Thus, the structure explains well how Rev1 incorporates dCMP opposite abasic sites.

Recently, the structure of human Rev1 catalytic core in complex with DNA and dCTP was shown to be very similar to that of the yeast Rev1 (Fig. 1H), with the exception of two insertions in hRev1 (Swan *et al.*, 2009). One insert (I1, ~40 residues of the 378–417 region) in the palm domain extends away from the active site and the other insert (I2, 54 residues of the 449–504 region) in the finger domain may constitute, together with the G-loop in the PAD, a hydrophobic pocket to accommodate the evicted template G with bulky adducts at N^2 position. While some portion of this I2 region is disordered in the complex with nondamaged DNA (Swan *et al.*, 2009), it may form a defined structure when complexed with DNA containing a bulky N^2–dG adduct. In fact, hRev1 was shown to bind DNA containing dG with CH_2-(6-benzo[*a*]pyrene) at N^2 position threefold more tightly than unmodified G-containing DNA (Choi and Guengerich, 2008). In the human Rev1–DNA–dCTP ternary complex, the G-loop in the PAD has multiple contacts with the αE helix in the I2 region of the finger domain, so as to encircle the hole through which the

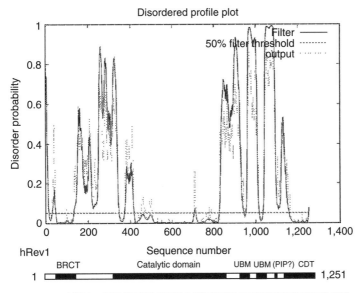

UBM : 935-SPSQLDQSVLEALPPDLREQVEQVCAVQQ-962
UBM : 1,012-AFSQVDPEVFAALPAELQRELKAAYDQRQ-1,040
(PIP?: 1,110-QKLIDGFL-1117)

FIG. 7. Disordered profile plot of hRev1. The location and sequence of a putative PIP-like sequence (1,110–1,117) in the C-terminal region is indicated by "PIP?". However, as shown in Fig 4A, a region of Rev1 spanning the putative PIP sequence (1,102–1,124) showed very weak PCNA binding activity in yeast two-hybrid assay.

template DNA strand extends its chain without any bending or kink. This implies that the interaction between the finger domain and PAD occurs after DNA binding; otherwise, longer chains of the template DNA cannot penetrate into such a hole. Thus, it is expected that the apo form of hRev1 should have a different conformation from the DNA-bound form.

From the disordered profile plot shown in Fig. 7, hRev1 is predicted to have a mosaic structure composed of multiple ordered and disordered regions. The ordered region near the N-terminus (50–130) corresponds to the BRCT domain. The catalytic domain (340–830) is flanked with long disordered regions. The two small peaks around 400 correspond to the I1 region that was disordered in the hRev1–DNA–dCTP ternary complex (Swan *et al.*, 2009).

In the case of hRev1, a PCNA-binding site has not been definitively identified as yet. Ross *et al.* (2005) localized a PCNA-binding site between 923 and 1,047 of hRev1 using a mammalian two-hybrid system. However, because two UBMs (934–962 and 1,012–1,040) were later found to exist within the same region, it is plausible that the putative PCNA-binding site in hRev1 might represent one or both of the UBMs. Since these workers could not detect an interaction between PCNA and hRev1 in a yeast two-hybrid assay, they suggested that the interaction might be indirect. Guo *et al.* (2006a) reported that mouse Rev1 (mRev1) could bind directly to PCNA, in a manner that depends on the functional BRCT domain. According to these authors, "yeast two-hybrid experiments demonstrated that the N-terminal half of mouse Rev1 protein interacts directly with PCNA (data not shown)." Furthermore, two truncated forms of HA-tagged mRev1 containing the 1–240 or 1–413 region were pulled down with a GST–PCNA fusion protein, but not with GST. However, using yeast two-hybrid assays as shown in Fig. 4, we have not observed differences in PCNA-binding activity between the intact form of hRev1 and a truncated hRev1 lacking the N-terminal BRCT domain (our unpublished observations). As noted in a recent review by Guo *et al.* (2009), it cannot be excluded that the BRCT-dependent interaction between mouse Rev1 and PCNA was indirectly mediated by other protein(s) bound to the BRCT domain. In this context, it is worthy to note that some BRCT domains are known to bind to DNA rather than to proteins. Kobayashi *et al.* (2006) showed that the BRCT domain in the N-terminal region of human RFC1 (the largest subunit of replication factor C, a clamp loader) is involved in binding to 5′-phosphorylated double-stranded (ds) DNA. The binding required a region N-terminal to the BRCT domain that was expected to form an α-helix. These workers noted that the N-terminal region of hRev1 aligns well with the BRCT region (BRCT domain plus N-terminal region) of hRFC1 and observed that the hRev1 BRCT region bound dsDNA (cited in the above paper). Clearly, further work is necessary to test whether hRev1 BRCT domain mediates direct or indirect interactions with PCNA.

The C-terminal region in the downstream of the catalytic domain of hRev1 is predicted to have multiple disordered regions. The two regions with low disorder probabilities located at around 950 and 1,020 correspond to two UBMs (934–962 and 1,012–1,040). Another region with low disorder probabilities is located at 1,110, where we found a sequence,

1110-QKL<u>I</u>DG<u>F</u>L-1117, which is similar to the PIP-box consensus sequence. We detected only a very weak signal for PCNA-binding activity of the sequence when the 1,102–1,124 sequence was examined by yeast two-hybrid assay (Fig. 4A). Moreover, we did not see any significant increase in PCNA-binding activity when we introduced L1117F substitution to change it to QKL<u>I</u>DG<u>FF</u>, a sequence perfectly matching the PIP-box consensus sequence (data not shown). It seems possible that some residues near or within the PIP-box-like sequence may negatively influence PCNA binding.

About 90 amino acids near the C-terminus (1,160–1,251) of hRev1 is predicted to be highly ordered, probably because the region is expected to have secondary structure containing multiple α-helices. The C-terminal region of hRev1 (hRev1-CTD) was first recognized to be important for the interaction with hRev7, the noncatalytic subunit of hPol ζ (Murakumo *et al.*, 2001). The interaction seemed to explain well how Pol ζ might function in conjunction with Rev1 for bypass of various DNA lesions. A stable 1:1 complex of hRev1 and hRev7 proteins, both of which were overproduced in *E. coli* cells, was purified through gel filtration, but the enzyme activity of hRev1 in the complex showed no significant difference from that of uncomplexed hRev1 (Masuda *et al.*, 2003). Subsequently, the Rev1-CTD was found to interact also with three other Y-family polymerases (Guo *et al.*, 2003; Ohashi *et al.*, 2004, Tissier *et al.*, 2004), thus suggesting that hRev1 plays a central role during *in vivo* TLS processes. More recently, the hRev1-CTD was found to recognize short (~10 amino acids) sequences containing FF, which are present in hPol κ, hPol ι and hPol η (Ohashi *et al.*, 2009). However, all the sequences containing FF were not recognized by hRev1-CTD; for example, the C-terminal PIP-box sequences in hPol κ and hPol η did not bind to hRev1-CTD. The role of interactions between hRev1 and other Y family polymerases is discussed in more detail in a later section.

Jansen and colleagues generated mutant mice containing a defined deletion of the BRCT domain in the *Rev1* gene or a completely defective *Rev1* gene (denoted as *Rev1*^B/B and *Rev1*^−/−, respectively) (Jansen *et al.*, 2005, 2006). While *Rev1*^B/B mice were healthy and displayed normal SHMs of immunoglobulin (Ig) genes, *Rev1*^−/− mice showed a transient growth retardation and strand-biased defect in C-to-G transversions in Ig genes, which most likely involved the Rev1-mediated dCMP incorporation opposite abasic lesions. Immortalized MEF lines from such mice were established and *Rev1*^B/B cells showed a milder UV sensitivity than *Rev1*^−/− cells, implicating multiple roles of Rev1 in intracellular TLS processes (Jansen *et al.*, 2009). Rev1-disrupted mutant of the chicken lymphocyte cell line DT40 also showed high sensitivities to UV and cisplatin (Okada *et al.*, 2005). When various constructs carrying the intact or mutant form of the human *REV1* gene were expressed in the *rev1* mutant of DT40, the full-length form (1–1,251), an N-terminal deletion mutant (333–1,251) lacking the BRCT domain and a catalytic mutant (D570A/E571A) fully corrected the sensitivities to UV and cisplatin, but a C-terminal deletion mutant (1–827) or a mutant carrying only the C-terminal portion (923–1,251) did not (Ross *et al.*, 2005). The results indicated that the C-terminal portion of Rev1 is necessary for effective tolerance of DNA damage in DT40 cells, but requiring some additional function(s) encoded by the central portion (other than the catalytic function), while the N-terminal portion including the BRCT domain was not essential for the tolerance mechanism.

III.   Functional Significance of Protein–Protein Interactions
Involving TLS DNA Polymerases

A.   *PCNA Binding and Formation of Nuclear Foci by TLS DNA
Polymerases in Genotoxin-Treated and Genotoxin-Untreated Cells*

PCNA interacts with a number of proteins involved in replication, repair, cell cycle, and other functions (for a recent review, see Moldovan *et al.*, 2007). Most of those proteins have a conserved sequence, PIP-box (Warbrick, 1998), which is often presented as Qxx(M,L,I)xxFF. The p66 subunit of human Pol δ has the sequence 456-QVSITGFFQRK-466 near the C-terminus (K466 is the last residue of the protein). However, neither hPol κ, hPol ι nor hPol η has a sequence completely matching the canonical sequence. Thus the conserved Gln is replaced with Met in hPol η's PIP2 or Lys in hPol ι and hPol κ's PIP2. Nevertheless, each of the Y-family polymerase PIP-box sequences gives a clearly positive signal for sequence-specific interactions with PCNA when examined by yeast two-hybrid assay (see Fig. 4A). To evaluate relative affinities of the non-canonical PIP-box sequences for PCNA, synthetic peptides containing each of PIP-box sequences were examined for quantitative measurements of binding to unmodified PCNA by surface plasmon resonance (SPR) (Hishiki *et al.*, 2009). As summarized in Table 1, the results indicate

TABLE 1
Estimated K_d Values for Binding to PCNA, Ubiquitin, and hRev1-CTD

	PCNA[a]	Ubiquitin	hRev1-CTD[b]
hPol η	0.40 (PIP2)	81[c]	13
hPol ι	0.39	180[d]	69
hPol κ	4.5 (PIP2+PLTH)	38±2[e]	7.6

All in μM.
[a] All obtained by SPR (Hishiki et al., 2009).
[b] All obtained by SPR (Ohashi et al., 2009).
[c] Obtained by NMR (Bomar et al., 2007).
[d] Obtained by NMR (Bienko et al., 2005).
[e] K_d value for UBZ of hPol κ is not available. The value obtained for the UBZ of Rad8 by SPR (Crosetto et al., 2008) is shown. The UBZ domain sequence of Rad18 is very similar to that of hPol κ.

that the PIP-box sequences of hPol η (PIP2) and hPol ι have a similar level of affinity for PCNA with estimated K_d (dissociation constant) values of 0.40 and 0.39 μM, respectively, and that the PIP2 of hPol κ has a much lower affinity. While the wild-type PIP2 of hPol κ showed a clearly positive signal for PCNA interaction by yeast two-hybrid assay (data not shown), significant levels of PCNA binding to the PIP2 of hPol κ by SPR were detected only with an extension of the sequence PLTH at the C-terminus to make the sequence similar to the PIP2 of hPol η. In SPR assays, the variant of hPol η's PIP2 lacking the C-terminal PLTH sequence did not show any signal for PCNA binding (Hishiki et al., 2009). However, the data shown in Fig. 4A clearly indicate that the PLTH-deleted sequence still retains PCNA-binding activity, demonstrating that yeast two-hybrid assay is more sensitive for detecting PCNA interaction than SPR.

Judging from K_d values obtained by SPR assays, the affinities of the PIP-box sequences for PCNA seem to be much stronger than those of UBDs (UBZ in hPol η and hPol κ or UBM in hPol ι and Rev1) for Ub (see Table I). For example, the K_d value in the interaction between free Ub and the UBZ of hPol η was estimated to be around 81 μM by isothermal titration calorimetry (ITC) (Bomar et al., 2007). Therefore, it seems likely that the PIP2 of hPol η contributes to the interactions between hPol η and mUb-PCNA more significantly than does the UBZ, thereby determining the binding specificity. For evaluation of the affinities for homotrimeric PCNA, it is desirable to re-estimate K_d values by methods other than SPR, for example, ITC. Structural analyses of the PCNA bound to the peptide containing each of the PIP-box sequences provided reasonable explanations about why the noncanonical PIP-box sequences have lower affinities than that of the canonical PIP-box sequence (Hishiki *et al.*, 2009).

Much of the information on *in vivo* functions of PIP-box and UBZ or UBM domains has been obtained from nuclear focus formation assays in genotoxin-treated cultured cells. Kannouche *et al.* (2001) described that Green Fluorescent Protein (GFP)-fused form of hPol η (GFP–hPol η) formed nuclear foci in an ~10–15% of MRC5 cells without any DNA-damaging treatment and that the frequency of the cells with such foci increased up to 80% after UV irradiation. Because the proportion of the cells with GFP–hPol η foci in nondamaged cells corresponded to that of the cells in S-phase and also because such foci colocalized with PCNA, they suggested that hPol η is associated with replication factories during S-phase in nondamaged cells. Subsequently, Kannouche *et al.* (2004) found that PCNA becomes monoubiquitinated in a variety of human cells following UV irradiation, similarly to previous findings in yeast (Hoege *et al.*, 2002), and described that hPol η interacted specifically with mUb-PCNA, but not unmodified PCNA *in vivo*. Furthermore, Bienko *et al.* (2005) showed that both UBZ and PIP motifs of hPol η were essential for the nuclear foci formation in UV-irradiated cells. However, it remains unanswered how hPol η formed nuclear foci colocalizing with PCNA (presumably unmodified) in nondamaged cells. A similar study for hPol κ showed that the frequency of the cells with nuclear foci of GFP–hPol κ was around 5% of MRC5 cells expressing the fusion proteins under nondamaged conditions and it increased up to at most 23% of the cells (Ogi *et al.*, 2005). As the C-terminal 119 residues of hPol η that contains UBZ, NLS (nuclear localization signal) and PIP were required for nuclear foci formation (Kannouche *et al.*, 2001), the C-terminal 97 residues of hPol κ containing one of the two UBZs, NLS and PIP were necessary and sufficient for foci formation induced by DNA damages. Similarly, nuclear foci formation of hPol ι in DNA-damaged cells required the intact PIP (Vidal *et al.*, 2004) and at least one of the two UBMs (Bienko *et al.*, 2005). These results are consistent with the notion that formation of hPol η, hPol ι or hPol κ, nuclear foci in DNA-damaged cells depends on both PIP and UBZ or UBM, probably for their stable binding to mUb-PCNA.

However, results of nuclear focus formation assays should be interpreted with caution, since the nature of nuclear foci is poorly understood. For instance, we do not know how many molecules of a protein in question must accumulate to be detected as foci or how many other different proteins are present in the same foci. Even if two proteins (e.g., hPol η and hPol ι) colocalize in foci, the two proteins are not necessarily interacting directly with each other in such foci. Potentially, colocalization may occur through binding to a common binding partner (e.g., PCNA). The results obtained by Gueranger *et al.* (2008) suggested that TLS by hPol η might occur without the accumulation of microscopically visible foci. These workers isolated a mutant of the Burkitt's lymphoma cell line BL2 by inactivating the gene coding for hPol η and observed that the hPol η-deficient mutant exhibited a UV-sensitive phenotype. The UV-sensitive phenotype was fully restored by transfection of an EGFP expression vector carrying the wild-type hPol η, but also by expressing a mutant of hPol η with the alteration of the C-terminal PIP-box sequence (701-MQTLESFF-708) to MATAESAA that did not form detectable nuclear foci. The result implied that the functional complementation and accumulation at nuclear foci are separable, but it does not rule out a possibility that hPol η has another PIP-box in addition to the C-terminal one. Acharya *et al.* (2008) noted the presence of an additional PIP-like sequence 437-STDITSFL-444 (named PIP1, and the C-terminal one was named PIP2), just C-terminal to the PAD (at a very similar position to that of the PIP-box in hPol ι). Their results showed that the PIP1 mutant (F443A, L444A) or the PIP2 mutant (F707A, F708A) conferred intermediate levels of UV resistance to XP-V cells and the PIP1 PIP2 double mutant was completely defective in imparting UV resistance to XP-V cells. They concluded that hPol η has two PCNA-binding PIP domains that can functionally substitute for one another. We are interested in comparing relative strength in PCNA-binding activity between the PIP1 and PIP2 of hPol η and examined both sequences for PCNA-binding by yeast two-hybrid assay. As shown in Fig. 4A, we could not detect any signal of PIP1 for PCNA interaction in yeast two-hybrid assay, while we could observe a very strong signal for PCNA interaction with PIP2. It seems that PCNA-binding activity of the PIP1 sequence is much weaker than that of the PIP2 sequence, while a possibility that the PIP1 requires some additional sequence for exhibiting its PCNA-binding activity cannot be ruled out.

Acharya *et al.* (2008) examined nuclear foci formation by coexpressing GFP–PCNA and FLAG–hPol η in MRC5 cells and observed that the D652A mutation in the UBZ domain, which inactivated Ub binding of hPol η (Bienko *et al.*, 2005), still retained the ability to form nuclear foci that colocalized with PCNA. However, Sabbioneda *et al.* (2009) argued against the interpretation, by showing that the eGFP–hPol η carrying the D652A mutation failed to form foci without overexpression of PCNA. Thus, requirements for nuclear foci formation are variable depending on experimental conditions examined.

The question of what elements are required for nuclear foci formation of hRev1 is also under debates. Tissier *et al.* (2004) reported that when expressed as a fusion with Yellow Fluorescent Protein (YFP), each of N-terminal half (1–730) and C-terminal half (730–1,251) of hRev1 formed nuclear foci in UV-irradiated cells, whereas Murakumo *et al.* (2006) reported that nuclear foci formation of GFP–hRev1 in UV-irradiated cells was observed with the C-terminal region (e.g., 826–1,251), but not with the N-terminal region (e.g., 1–825). Furthermore, Guo *et al.* (2006a) reported that the intact BRCT domain was required for foci formation of GFP–mRev1 in nondamaged cells, but not in UV-irradiated cells. They also showed that UBMs are required for increased level of foci formation of GFP–mRev1 in UV-irradiated cells (Guo *et al.*, 2006b). If we assume that the C-terminal region of hRev1 has a weak PCNA-binding site as well as two UBMs, we can interpret the above results as indicating that requirements for nuclear foci formation of the C-terminal region of hRev1 in DNA-damaged cells are similar to those of hPol η, hPol κ, and hPol ι.

Also, it should be noted that all the above results were obtained with ectopically expressed human or mouse Rev1, but not with the endogenous level of the Rev1 proteins. Akagi *et al.* (2009) studied accumulation of the endogenous hRev1 into locally UV-irradiated areas of nuclei, using antibody with high affinity for hRev1, because nuclear foci of the endogenous hRev1 were not clearly observed even after UV irradiation. Importantly, they observed the accumulations of the endogenous hRev1 into locally UV-irradiated areas of nuclei in the cells expressing hPol η, but not in hPol η-deficient XP-V cells. This makes a sharp contrast to the observation by Tissier *et al.* (2004) that GFP–hRev1 formed nuclear foci in UV-irradiated XP-V cells. Furthermore, when XP-V cells were reconstituted with wild-type hPol η or its mutant defective for the hRev1 interaction, UV sensitivity of the XP-V cells was corrected by either the wild type or the mutant, yet the accumulation of the endogenous hRev1 into UV-irradiated areas was observed with the wild type, but not with the mutant. These results indicate that accumulation of the endogenous hRev1 to UV-irradiated areas depends on the interaction with hPol η, while nuclear foci formation of ectopically (over) expressed hRev1 occurred independently of hPol η. Thus, there are two different pathways for targeting hRev1 to sites of DNA damage, hPol η-dependent and hPol η-independent ones. The latter becomes more easily detectable when ectopically (over)expressed.

B. Mechanism and Biological Significance of Rev1–TLS Polymerase Interactions

Sequences as short as 10 aa residues in hPol κ, hPol ι, and hPol η are sufficient for mediating interaction with hRev1-CTD (Ohashi et al., 2009). Thus far, only four such sequences have been identified, all of which contain two consecutive phenylalanines (FF) as critical residues. As the PIP-box consensus sequence is often denoted as Qxx(I,L,M)xxFF, many PIP-box sequences contain FF. However, none of the four RIR sequences binds to PCNA and the PIP-box of hPol ι did not bind to hRev1-CTD (Fig. 4B). The PIP-box sequence containing FF at the C-terminus of hPol η or hPol κ did not bind to hRev1-CTD (Ohashi et al., 2009). In PIP-box sequences, several residues in the N-terminal side of FF are required for interaction with multiple residues in PCNA and adopting a 3_{10} helical structure, while residues C-terminal to FF are not essential although they contribute to stabilizing the interaction with PCNA. By contrast, in RIR sequences, no conserved aa is present in either N-terminus or C-terminus to FF; however, the presence of several sequences in the C-terminal side of FF is essential for binding to hRev1-CTD. Proline substitution of the four residues C-terminal to FF abrogated the hRev1-CTD interaction, while alanine substitution did not. Thus, the consensus sequence for RIR is denoted as xxxFFyyyy (y, any residue but not proline). For evaluating relative affinities of the RIRs in hPol η, hPol ι, and hPol κ, synthetic peptides carrying each of the RIR sequences were examined for binding to His-hRev1(1,130–1,251) by SPR. As summarized in Table 1, the RIR of hPol κ showed higher affinity than that of hPol η or hPol ι. The hRev1-CTD is known to bind to hRev7, which contains no FF sequence in the entire sequence of 211 residues. Moreover, a longer region (>150 residues) of hRev7 is required for binding to hRev1-CTD (Murakumo *et al.*, 2001), while hPol η, hPol ι, and hPol κ interact with hRev1-CTD via short RIR sequences. It therefore seems likely that hRev1-CTD has two different interfaces for interactions with other proteins, one recognizing short RIR sequence in hPol η, hPol ι, and hPol κ and the other recognizing conformation of hRev7. How hRev1-CTD recognizes the RIR sequences is an intriguing question, but has not been solved as yet because of the difficulty in purifying the hRev1-CTD at large scales for structural analysis.

The *Saccharomyces cerevisiae* (sc) proteins involved in TLS are known to show physical interactions, while some differences exist in such interactions among yeast and human proteins. The scRev7 protein interacts mainly with the PAD domain of scRev1 (Acharya *et al.*, 2005), also interacting weakly with the CTD and BRCT domains (D'Souza and Walker, 2006). The PAD of scRev1 interacts with a C-terminal region of scPol η (Acharya *et al.*, 2007) and the scRev1-CTD interacted with a central region of scREV3 (Acharya *et al.*, 2006). More recently, interactions among the Y-family proteins in the fruit fly *Drosophila melanogaster* (dm) were reported (Kosarek *et al.*, 2008). The C-terminal 117 amino acids of dmRev1 were necessary and sufficient for an interaction with dmPol η, but a region adjacent to the C-terminus of dmRev1 was required for its interaction with dmPol ι. Interestingly, dmPol η, but not dmPol ι, interacted with the C-terminal region (~100 residues) of mRev1. Since the C-terminal sequences of the human and mouse Rev1 proteins are well conserved with 95% identity (Masuda *et al.*, 2002), the mRev1-CTD is expected to recognize the same RIR sequences as the hRev1-CTD. While dmPol ι has no FF in the entire sequence, dmPol η has five sites containing FF (26–27, 500–501, 564–565, 786–787, 880–881). A mutant of dmPol η with FF26–27AA and FF564–565AA substitutions was found to have lost most of the dmRev1-interacting activity (J. Tomida and T. Todo, personal communication). This suggests that the dmRev1-CTD, which is distantly related to the hRev1- and mRev1-CTDs (24% identity), recognizes sequences containing FF, while at the present it is not explainable why three other sites containing FF do not interact with dmRev1-CTD. In any case, interactions of Rev1 with other TLS polymerases appear to be conserved in eukaryotes with some variations.

The most important unanswered question about Rev1 concerns its central role during the intracellular TLS processes. As described above, a hPol κ mutant defective for interaction with hRev1 could not correct the

BPDE sensitivity of *Polk*-defective MEF cells (Ohashi *et al.*, 2009), implying that the hRev1 interaction is essential for hPol κ to execute its function *in vivo*. In contrast, a similar mutant of hPol η defective for hRev1 interaction corrected the two phenotypes of XP-V cells, namely, UV sensitivity and elevated mutation rates by UV irradiation as efficiently as the wild type (Akagi *et al.*, 2009). Nevertheless, the hPol η mutant suppressed only partially another phenotype of XP-V cells, that is, a higher incidence of spontaneous mutations, while the wild type suppressed the phenotype completely. This suggested that, while the Rev1 interaction is dispensable for hPol η to carry out accurate TLS of UV-induced lesions such as CPDs, it contributes to suppression of spontaneous mutations, probably by promoting accurate TLS past endogenous DNA lesions such as those generated by oxidative stress. While no data are available at the present to infer whether or not the Rev1 interaction is essential for hPol ι function, hPol ι is known to interact with hPol η and relocalization of GFP–hPol ι in foci after UV irradiation was shown to depend on hPol η (Kannouche *et al.*, 2002). It is perhaps surprising that Pol ι recruitment is dependent on Pol η since hPol ι has a PIP-box with affinity for PCNA similar to that of hPol η's PIP2 (Hishiki *et al.*, 2009) and two UBMs for binding to mUb-PCNA (Bienko *et al.*, 2005). These results that recruitment of the endogenous hRev1 or hPol ι to DNA-damaged sites is dependent on hPol η cannot be explained by a model which presumes that TLS polymerases are recruited to DNA-damaged sites randomly on a simple "try and error" basis.

We therefore proposed a "sequential recruitment" model to explain the mechanism by which TLS polymerases are recruited to stalled replication forks at sites of DNA damage (Barkley *et al.*, 2007). When the replication fork encounters a site of DNA damage, uncoupling between the preceding DNA helicase and the stalled replicative DNA polymerase generates single-stranded (ss) regions on the template DNA, to which RPA (a heterotrimeric protein with ssDNA-binding activity) binds. The RPA-coated ssDNA region then recruits the Rad6 (an E2 Ub-conjugating enzyme)–Rad18 (an E3 Ub ligase) complex via the interaction between RPA and Rad18 (Davies *et al.*, 2008; Tsuji *et al.*, 2008) to mediate monoubiquitination of PCNA in the stalled replication fork, which facilitates transfer of hPol η from the Rad6–Rad18 complex to mUb-PCNA. Preferential recruitment of hPol η to all stalled forks could be advantageous for cells because hPol η is a versatile enzyme capable of by passing many different DNA lesions correctly in many cases (Masutani *et al.*, 2000). However, hPol η forms nuclear foci in response to various DNA-damaging treatments, including BPDE that mainly generates N^2-BPDE–dG adducts which the enzyme cannot bypass (Bi *et al.*, 2005). Therefore, the hPol η recruited in response to BPDE lesions needs to be replaced with hPol κ that is able to bypass N^2-BPDE–dG adducts correctly (Ogi *et al.*, 2002). Such a polymerase switching might be dependent on hRev1, which interacts with both hPol η and hPol κ. A similar scenario can be envisaged for other lesions to explain the exchange between two TLS polymerases, one for inserting a nucleotide opposite a lesion and the other for extending further. For example, at the site of T–T (6–4)PP, hPol η inserts one of the 4XMPs opposite 3′-T, but does not extend further (Masutani *et al.*, 1999a). If hPol η is replaced with hPol ι before inserting any nucleotide opposite the 3′-T (either via direct hPol η–hPol ι interactions, or through interactions of hPol ι and hPol η with hRev1), hPol ι can preferentially insert the correct A opposite the 3′-T without extending further. In both cases, hPol ζ is believed to carry out the extension reaction, after being recruited most probably through the interaction between hRev1 and hRev7, the noncatalytic subunit of hPol ζ.

Another model was proposed, which postulates that TLS enzymes might be recruited by two different mechanisms, depending on the coding capacity of the damaged nucleotide at the template strand (Jansen *et al.*, 2007). First, damaged nucleotides retaining good coding capacity (e.g., CPD) may be readily bypassed by hPol η, without conferring a significant replication arrest. Second, forms of damage causing severe distortion of the DNA structure and/or having poor coding capacity [e.g., (6–4)PP] activates a more elaborate pathway that requires functioning of multiple TLS polymerases for the insertion and extension steps separately, repriming of replication downstream of the lesion and activation of DNA damage signaling which involves ATR and an alternative clamp, the Rad9–Rad1–Hus1 (9–1–1) complex. Further according to this model, the 9–1–1 clamp binds to the reprimed 5′ terminus and recruits the hRev1/hPol ζ complex, which then moves up to the stalled 3′ terminus by the unique activity of hRev1 to bind to stretches of ssDNA and then translocate to 3′ primer–template junctions (Masuda and Kamiya, 2006). The authors also consider that hPol η is the default primary TLS polymerase recruited to a stalled fork and that hRev1 may dislodge the defunct hPol η from the stalled 3′ terminus. It may be also possible that hRev1 is recruited to the reprimed 5′ terminus, either as the hRev1/Pol ζ complex via a presumed interaction between hRev7 and the 9–1–1 complex or by itself through binding of its N-terminal BRCT domain to 5′ recessed end of dsDNA (as discussed earlier).

Thus far, there are two reports on physical interactions between TLS proteins and the 9–1–1 complex, both from studies on yeasts; in S. cerevisiae, Rev7 binds to the Hus1 and Rad9 orthologs (Sabbioneda et al., 2005) and in S. pombe, DinB/Pol κ interacts with the 9–1–1 complex, especially with Hus1 (Kai and Wang, 2003). However, at the moment, it is not known whether such interactions occur among the mammalian counterparts. Very recently, the structure of the human 9–1–1 complex has been determined (Dore et al., 2009; Sohn and Cho, 2009; Xu et al., 2009), which revealed that the 9–1–1 complex has a structure very similar to that of PCNA. It should be very intriguing to examine whether PIP-box of any human TLS polymerase (and hREV7) can bind to the human 9–1–1 complex or its subunit and if it does bind, how strongly it binds in comparison with PCNA binding.

As pertains to the two different pathways for recruitment of TLS proteins to sites of DNA damage, one may argue that the replication block should be more severe when the replication fork encounters a blocking DNA damage present on the leading strand template than on the lagging strand template, because repriming occurs constantly for the lagging strand synthesis of a new Okazaki fragment, even in the absence of DNA damage. Therefore, it may be plausible that different pathways for TLS may become activated, depending on which strand a blocking lesion is present. In contrast to the models described above, Edmunds et al. (2008) proposed a different model in which Rev1 functions for TLS at a stalled replication fork independently of PCNA ubiquitination and that Rad18-dependent PCNA ubiquitination controls TLS at a postreplicative gap filling. These authors proposed that Rev1 might be recruited to the stalled fork via interaction with the replicative DNA polymerase and/or accessory proteins and then recruits another TLS polymerase, such as Pol η. In this connection, it is noteworthy that scRev1 protein interacts via its PAD with the Pol32 protein, which is the nonessential subunit of scPol δ involved in mutagenesis (Acharya et al., 2009). Previous site-specific mutagenesis experiments using plasmid DNA containing an abasic site, T–T CPD or T–T (6–4)PP demonstrated that Pol32 is required for the bypass of abasic sites and T–T (6–4)PP in a manner that is dependent on scRev1 and scPol ζ, but not for the bypass of T–T CPD which is scPol η dependent (Gibbs *et al.*, 2005). In contrast, the *rev6-1* mutation corresponding to the G178S substitution in PCNA abolishes the bypass of all three lesions (Zhang *et al.*, 2006). Because Pol32 binds to the scRev1–scPol ζ complex via its central region, binding to the Pol31 subunit of scPol δ via its N-terminal region and to PCNA via the C-terminal PIP-box (Acharya *et al.*, 2009), it follows that the scRev1–scPol ζ complex can bind to the scPol δ bound to PCNA in the replisome. However, even if such an interaction between Pol δ and the Rev1/Pol ζ occurs in human cells, it does not explain why relocalization of the endogenous hRev1 after DNA damaging is dependent on hPol η.

C. *Interactions with Other Proteins and Posttranslational Modifications*

There are many reports describing that the activities of hPol η and other TLS polymerases are stimulated by the addition of various proteins such as Msh2 (Wilson *et al.*, 2005), or WRN (Werner syndrome protein, Kamath-Loeb *et al.*, 2007) and Ctf18–Replication factor C complex (Shiomi *et al.*, 2007). However, since no mutant specifically affecting such interactions is available at the present, we cannot speculate on the biological relevance of such interactions and we must wait for more detailed results of further experimentations. Rather, we would like to point out that disordered regions of a TLS protein or any given protein have much more potentials for interaction with other proteins than ordered regions, in which most of the sequences in ordered regions are employed for the maintenance of specific secondary structures and only limited sequences of them are available for interactions with other proteins. Interactions of short motif sequences in disordered regions with other proteins may be mostly weak and transient ones, involving dynamic exchanges of binding partners. Such interactions should be important *in vivo* during TLS because TLS is an inherently transient process, acting only during the period when replicative DNA polymerases are stalled. It is also known that unstructured regions are susceptible to various posttranslational modifications, such as phosphorylation (Fink, 2005). A recent paper suggested that hPol η might be phosphorylated at S587 and T617, both of which are located in the C-terminal unstructured region (Chen *et al.*, 2008). It is likely that future studies will identify additional posttranslational modifications of Y-family polymerases as important regulatory mechanisms for TLS.

Acknowledgments

We thank Katsuhiko Sasaki for his excellent technical assistance and also Junya Tomida and Takeshi Todo for permitting us to cite their unpublished results. Our works described and cited here were supported by grants-in-aids (17013041 to H. O.) from the Ministry of Education, Sports, Science, and Technology of Japan and by grants (ES09558 to C. V.) from the National Institutes of Health, USA.

References

Acharya, N., Haracska, L., Johnson, R. E., Unk, I., Prakash, S., and Prakash, L. (2005). Complex formation of yeast Rev1 and Rev7 proteins: A novel role for the polymerase-associated domain. *Mol. Cell. Biol.* **25**, 9734–9740.

Acharya, N., Haracska, L., Prakash, S., and Prakash, L. (2007). Complex formation of yeast Rev1 with DNA polymerase η. *Mol. Cell. Biol.* **27**, 8401–8408.

Acharya, N., Johnson, R. E., Pages, V., Prakash, S., and Prakash, L. (2009). Yeast Rev1 protein promotes complex formation of DNA polymerase ζ with P32 subunit of DNA polymerase δ. *Proc. Natl. Acad. Sci. USA* **106**, 9631–9636.

Acharya, N., Johnson, R. E., Prakash, S., and Prakash, L. (2006). Complex formation with Rev1 enhances the proficiency of *Saccharomyces cerevisiae* DNA polymerase ζ for mismatch extension and for extension opposite from DNA lesions. *Mol. Cell. Biol.* **26**, 9555–9563.

Acharya, N., Yoon, J.-H., Gali, H., Unk, I., Haracska, L., Johnson, R. E., Hurwitz, J., Prakash, L., and Prakash, S. (2008). Roles of PCNA-binding and ubiquitin-binding domains in human DNA polymerase η in translesion DNA synthesis. *Proc. Natl. Acad. Sci. USA* **105**, 17724–17729.

Akagi, J., Masutani, C., Kataoka, Y., Kan, T., Ohashi, E., Mori, T., Ohmori, H., and Hanaoka, F. (2009). Interaction with DNA polymerase η is required for nuclear accumulation of REV1 and suppression of spontaneous mutations in human cells. *DNA Repair* **8**, 585–599.

Alt, A., Lammens, K., Chiocchini, C., Lammens, A., Pieck, J. C., Kuch, D., Hopner, K. -P., and Carell, T. (2007). Bypass of DNA lesions generated during anticancer treatment with cisplatin by DNA polymerase η. *Science* **318**, 967–970.

Auffret van der Kemp, P., de Padula, M., Burguiere-Slezak, G., Ulrich, H. D., and Boiteux, S. (2009). PCNA monoubiquitylation and DNA polymerase η ubiquitin-binding domain are required to prevent 8-oxoguanine-induced mutagenesis in *Saccharomyces cerevisiae*. *Nucleic Acids Res.* **37**, 2549–2559.

Barkley, L. R., Ohmori, H., and Vaziri, C. (2007). Integrating S -phase checkpoint signaling with trans-lesion synthesis of bulky DNA adducts. *Cell Biochem. Biophys.* **47**, 392–408.

Bi, X., Slater, D. M., Ohmori, H., and Vaziri, C. (2005). DNA polymerase κ is specifically required for recovery from the benzo[*a*]pyrene-dihydrodiol epoxide (BPDE)-induced S-phase checkpoint. *J Biol. Chem.* **280**, 22343–22355.

Bienko, M., Green, C. M., Crosetto, N., Rudolf, F., Zapart, G., Coull, B., Kannouche, P., Wider, G., Peter, M., Lehmann, A. R., Hofmann, K., and Dikic, I. (2005). Ubiquitin-binding domains in Y-family polymerases regulate translesion synthesis. *Science* **310**, 1821–1824.

Bomar, M. G., Pai, M.-T., Tzeng, S.-R., Li, S. S.-C., and Zhou, P. (2007). Structure of the ubiquitin -binding zinc finger domain of human DNA Y-polymerase η. *EMBO Rep.* **8**, 247–251.

Boudsocq, F., Kokoska, R. J., Plosky, B. S., Vaisman, A., Ling, H., Kunkel, T. A., Yang, W., and Woodgate, R. (2004). Investigating the role of the little finger domain of Y-family DNA polymerases in low fidelity synthesis and translesion replication. *J. Biol. Chem.* **279**, 32932–32940.

Carpio, R. V.-D., Silverstein, T. D., Lone, S., Swan, M. K., Choudhury, J. R., Johnson, R. E., Prakash, S., Prakash, L., and Aggarwal, A. K. (2009). Structure of human DNA polymerase κ inserting dATP opposite an 8-oxoG DNA lesion. *PLoS One* **4**, e5766.

Chen, Y.-W., Cleaver, J. E., Hatahet, Z., Honkamen, R. E., Chang, J.-Y., Yen, Y., and Chou, K.-M. (2008). Human DNA polymerase h activity and translocation is regulated by phosphorylaiton. *Proc. Natl. Acad. Sci. USA* **105**, 16578–16583.

Choi, J.-Y., Angel, K. C., and Guengerich, F. P. (2006). Translesion synthesis across bulky N^2-alkyl guanine DNA adducts by human DNA polymerase κ. *J. Biol. Chem.* **281**, 21062–21072.

Choi, J.-Y., and Guengerich, F. P. (2008). Kinetic analysis of translesion synthesis opposite bulky N^2- and O^6-alkylguanine DNA adducts by human DNA polymerase REV1. *J. Biol. Chem.* **283**, 23645–23655.

Crosetto, N., Bienko, M., Hibbert, R. G., Perica, T., Ambrogio, C., Kensche, T., Hofmann, K., Sixma, T. K., and Dikic, I. (2008). Human Wrnip1 is localized in replication factories in a ubiquitin-binding zinc finger dependent manner. *J. Biol. Chem.* **283**, 35173–35185.

Davies, A. A., Huttner, D., Daigaku, Y., Chen, S., and Ulrich, H. D. (2008). Activation of ubiquitin-dependent DNA damage bypass is mediated by replication protein A. *Mol. Cell* **29**, 625–636.

de Padula, M., Slezak, G., Auffret van Der Kemp, P., and Boiteux, S. (2004). The post-replication repair RAD18 and RAD6 genes are involved in the prevention of spontaneous mutations caused by 7,8-dihydro-8-oxoguanine in *Saccharomyces cerevisiae*. *Nucleic Acids Res.* **23**, 5003–5010.

Dore, A. S., Kilkenny, M., Rzechorzek , N. J., and Pearl, L. H. (2009). Crystal structure of the human Rad9-Rad1-Hus1 DNA damage checkpoint complex-Implications for clamp loading and regulation. *Mol. Cell* **34**, 735–745.

D'Souza, S., and Walker, G. C. (2006). Novel role for the C terminus of *Saccharomyces cerevisiae* mediating protein-protein interactions. *Mol. Cell. Biol.* **26**, 8173–8182.

Dunker, A. K., Silman, I., Uversky, V. N., and Sussman, J. L. (2008). Function and structure of inherently disordered proteins. *Curr. Opin. Struct. Biol.* **18**, 756–764.

Edmunds, C. E., Simpson, L. J., and Sale, J. E. (2008). PCNA ubiquitination and REV1 define temporally distinct mechanisms for controlling translesion synthesis in the avian cell line DT40. *Mol. Cell* **30**, 519–529.

Fink, A. L. (2005). Natively unfolded proteins. *Curr. Opin. Struct. Bol.* **15**, 35–41.

Friedberg, E. C., Walker, G. C., Siede, W., Wood, R. D., Shultz, R. A., and Ellenberger, T. (2006). "DNA Repair and Mutagenesis." 2nd edn., ASM Press, Washington, DC.

Gibbs, P. E.M., Wang, X.-D., Li, Z., McManus, T. P., McGregor, W. G., Lawrence, C. W., and Maher, V. M. (2000). The function of the human homolog of *Saccharomyces cerevisiae REV1* is required for mutagenesis induced by UV light. *Proc. Natl. Acad. Sci. USA* **97**, 4186–4191.

Gibbs, P. E.M., McDonald, J., Woodgate, R., and Lawrence, C. W. (2005). The relative roles *in vivo* of *Saccharomyces cerevisiae* Pol η, Pol ζ, Rev1 protein and Pol32 in the bypass and mutation induction of an abasic site, T-T (6-4) photoadduct and T-T *cis-syn* cyclobutane dimer. *Genetics* **169**, 575–582.

Gueranger, Q., Stary, A., Aoufouchi, S., Fali, A., Sarasin, A., Reynaud, C.-A., and Weill, J.-C. (2008). Role of DNA polymerases η, ι and ζ in UV resistance and UV-induced mutagenesis in a human cell line. *DNA Repair* **7**, 1551–1562.

Guo, C., Fischhaber, P. L., Luk-Paszyc, M. J., Masuda, Y., Zhou, J., Kamiya, K., Kisker, C., and Friedberg, E. C. (2003). Mouse Rev1 protein interacts with multiple DNA polymerases involved in translesion DNA synthesis. *EMBO J.* **22**, 6621–6630.

Guo, C., Kosarek-Stancel, J. N., Tang, T. S., and Friedberg, E. C. (2009). Y-family DNA polymerases in mammalian cells. *Cell. Mol. Life Sci.* **66**, 2363–2381.

Guo, C., Sonoda, E., Tang, T. S., Parker, J. L., Bielen, A. B., Takeda, S., Ulrich, H. D., and Friedberg, E. C. (2006a). REV1 protein interacts with PCNA: Significance of the REV1 BRCT domain in vitro and in vivo. *Mol. Cell* **23**, 265–271.

Guo, C., Tang, T. S., Bienko, M., Parker, J. L., Bielen, A. B., Sonoda, E., Takeda, S., Ulrich, H. D., Dikic, I., and Friedberg, E. C. (2006b). Ubiquitin-binding motifs in REV1 protein are required for its role in the tolerance of DNA damage. *Mol. Cell. Biol.* **26**, 8892–8900.

Haracska, L., Acharya, N., Unk, I., Johnson, R. E., Hurwitz, J., Prakash, L., and Prakash, S. (2005). A single domain in human DNA polymerase ι mediates interaction with PCNA: Implications for translesion DNA synthesis. *Mol. Cell. Biol.* **25**, 1183–1190.

Haracska, L., Johnson, R. E., Unk, I., Phillips, B. B., Hurwitz, J., Prakash, L., and Prakash, S. (2001). Targeting of human DNA polymerase ι to the replication machinery via interaction with PCNA. *Proc. Natl. Acad. Sci. USA* **98**, 14256–14261.

Haracska, L., Prakash, S., and Prakash, L. (2002). Role of human DNA polymerase κ as an extender in translesion synthesis. *Proc. Natl. Acad. Sci. USA* **99**, 16000–16005.

Haracska, L., Yu, S. L., Johnson, R. E., Prakash, L., and Prakash, S. (2000). Efficient and accurate replication in the presence of 7,8-dihydro -8-oxoguanine by DNA polymerase η. *Nat. Genet.* **25**, 458–461.

Hishiki, M., Hashimoto, H., Hanafusa, T., Kamei, K., Ohashi, E., Shimizu, T., Ohmori, H., and Sato, M. (2009). Structural basis for novel interactions between human translesion synthesis polymerases and proliferating cell nuclear antigen. *J. Biol. Chem.* **284**, 110552–10560.

Hoege, C., Pfander, B., Moldova, G. L., Pyrowolakis, G., and Jentsch, S. (2002). RAD6-dependent DNA repair is linked to modification of PCNA by ubiquitin and SUMO. *Nature* **419**, 135–141.

Irimia, A., Eoff, R. L., Guengerich, F. P., and Egli, M. (2009). Structural and functional elucidation of the mechanism promoting error-prone synthesis by human DNA polymerase κ opposite 7,8-dihydro-8-oxo-2′-deoxyguanosine adduct. *J. Biol. Chem.*, **284**, 22467–22480.

Jain, R., Nair, D. T., Johnson, R. E., Prakash, L., Prakash, S., and Aggarwal, A. K. (2009). Replication across template T/U by human DNA polymerase-ι. *Structure* **17**, 974–980.

Jaloszynski, P., Ohashi, E., Ohmori, H., and Nishimura, S. (2005). Error-prone and inefficient replication across 8-hydroxyguanine (8-oxoguanine) in human and mouse *ras* gene fragments by DNA polymerase κ. *Genes Cells* **10**, 543–550.

Jansen, J. G., Fousteri, M. I., and de Wind, N. (2007). Send in the clamps: Control of DNA translesion synthesis in eukaryotes. *Mol. Cell* **28**, 522–529.

Jansen, J. G., Langerak, P., Tsaalbi-Shtylik, A., van den Berk, P., Jacobs, H., and de Wind, N. (2006). Strand-biased defect in C/G transversions in hypermutating immunoglobulin genes in Rev1-deficient mice. *J. Exp. Med.* **203**, 319–323.

Jansen, J. G., Tsaalbi-Shtylik, A., Langerak, P., Calléja, F., Meijers, C. M., Jacobs, H., and de Wind, N. (2005). The BRCT domain of mammalian Rev1 is involved in regulating DNA translesion synthesis. *Nucleic Acids Res.* **33**, 356–365.

Jansen, J. G., Tsaalbi-Shtylik, A., Hendriks, G., Gali, H., Hendel, A., Johansson, F., Erixon, K., Livneh, Z., Mullenders, L. H., Haracska, L., and de Wind, N. (2009). Separate domains of Rev1 mediate two modes of DNA damage bypass in mammalian cells. *Mol. Cell. Biol.* **29**, 3113–3123.

Jarosz, D. F., Godoy, V. G., Delaney, J. C., Essigmann, J. M., and Walker, G. C. (2006). A single amino acid governs enhanced activity of DinB DNA polymerases on damaged templates. *Nature* **439**, 225–228.

Jia, L., Geacintov, N. E., and Broyde, S. (2008). The N-claps of human DNA polymerase κ promotes blockage or error-free bypass of adenine- or guanine-benzo[a]pyrenyl lesions. *Nucleic Acids Res.* **36**, 6571–6584.

Johnson, R. E., Prakash, S., and Prakash, L. (1999a). Efficient bypass of a thymine-thymine dimer by yeast DNA polymerase, Polη. *Science* **283**, 1001–1004.

Johnson, R. E., Kondratick, C. M., Prakash, S., and Prakash, L. (1999b). hRAD30 mutations in the variant form of xeroderma pigmentosum. *Science* **285**, 263–265.

Kai, M., and Wang, T. S. (2003). Checkpoint activation regulates mutagenic translesion synthesis. *Genes Dev.* **17**, 64–76.

Kamath-Loeb, A. S., Lan, Li., Nakajima, S., Yasui, A., and Loeb, L. A. (2007). Werner syndrome protein interacts functionally with translesion DNA polymerases. *Proc. Natl. Acad. Sci. USA* **104**, 10394–10399.

Kannouche, P., Broughton, B. C., Volker, M., Hanaoka, F., Mullenders, L. H., and Lehmann, A. R. (2001). Domain structure, localization, and function of DNA polymerase η, defective in xeroderma pigmentosum variant cells. *Genes Dev.* **15**, 158–172.

Kannouche, P., Fernandez de Henestrosa, A. R., Coull, B., Vidal, A., Gray, C., Zicha, D., Woodgate, R., and Lehmann, A. R. (2002). Localization of DNA polymerase η and ι to the replication machinery is tightly coordinated in human cells. *EMBO J.* **21**, 6246–6256.

Kannouche, P., Wing, J., and Lehmann, A. R. (2004). Interaction of human DNA polymerase η with monoubiquitinated PCNA; A possible mechanism for the polymerase switch defective in response to DNA damage. *Mol. Cell* **14**, 491–500.

Kim, S. R., Maenhaut-Michel, G., Yamada, M., Yamamoto, Y., Matsui, K., Sofuni, T., Nohmi, N., and Ohmori, H. (1997). Multiple pathways for SOS-induced mutagenesis in *Escherichia coli*: An overproduction of *dinB/dinP* results in strongly enhancing mutagenesis in the absence of any exogenous treatment to damage DNA. *Proc. Natl. Acad. Sci. USA* **94**, 13792–13797.

Kirouac, K. N., and Ling, H. (2009). Structural basis of error-prone replication and stalling at a thymine base by human DNA polymerase ι. *EMBO J.* **28**, 1644–1654.

Kobayashi, M., Figaroa, F., Meeuwenoord, N., Jansen, L. E., and Siegal, G. (2006). Characterization of the DNA binding and structural properties of the BRCT region of human replication factor C p140 subunit. *J. Biol. Chem.* **281**, 4308–4317.

Kornberg, A., and Baker, T. (1991). "DNA Replication." 2nd edn., W. H. Freeman and Company, New York.

Kosarek, J. N., Woodruff, R. V., Rivera-Begeman, A., Guo, C., D'Souza, S., Koonin, E. V., Walker, G. C., and Friedberg, E. C. (2008). Comparative analysis of in vivo interactions between Rev1 protein and other Y-family DNA polymerases in animals and yeasts. *DNA Repair* **7**, 439–451.

Kulaeva, O. I., Koonin, E. V., McDonald, J. P., Randall, S. K., Rabinovich, N., Connaughton, J. F., Levine, A. S., and Woodgate, R. (1996). Identification of a DinB/UmuC homolog in the archeon *Sulfolobus solfataricus*. *Mutat. Res.* **357**, 245–253.

Lawrence, C. W. (2004). Cellular Functions of DNA polymerase ζ and Rev1 protein. *Adv. Prot. Chem.* **69**, 167–203.

Ling, H., Boudsocq, F., Woodgate, R., and Yang, W. (2001). Crystal structure of a Y-family DNA polymerase in action: A mechanism for error-prone and lesion-bypass replication. *Cell* **107**, 91–102.

Lone, S., Townson, S. A., Uljoin, S. N., Johnson, R. E., Brahma, A., Nair, D. T., Prakash, S., Prakash, L., and Aggarwal, A. K. (2007). Human DNA polymerase κ encircles DNA: Implications for mismatch extension and lesion bypass. *Mol. Cell* **25**, 601–614.

Masuda, Y., and Kamiya, K. (2006). Role of single-stranded DNA in targeting REV1 to primer termini. *J. Biol. Chem.* **281**, 24313–24321.

Masuda, Y., Ohmae, M., Masuda, K., and Kamiya, K. (2003). Structure and enzymatic properties of a stable complex of the human REV1 and REV7 proteins. *J. Biol. Chem.* **278**, 12356–12360.

Masuda, Y., Takahashi, M., Fukuda, S., Sumii, M., and Kamiya, K. (2002). Mechanism of dCMP transferase reactions catalyzed of mouse Rev1 protein. *J. Biol. Chem.* **277**, 3040–3046.

Masuda, Y., Takahashi, M., Tsunekuni, N., Minami, T., Sumii, M., Miyagawa, K., and Kamiya, K. (2001). Deoxycytidyl transferase activity of the human REV1 protein is closely associated with the conserved polymerase domain. *J. Biol. Chem.* **276**, 15051–15058.

Masutani, C., Araki, M., Yamada, A., Kusumoto, R., Nogimori, T., Maekawa, T., Iwai, S., and Hanaoka, F. (1999a). Xeroderma pigmentosum variant (XP-V) correcting protein from HeLa cells has a thymine dimer bypass DNA polymerase activity. *EMBO J.* **18**, 3491–3501.

Masutani, C., Kusumoto, R., Iwai, S., and Hanaoka, F. (2000). Mechanisms of accurate translesion synthesis by human DNA polymerase η. *EMBO J.* **19**, 3100–3109.

Masutani, C., Kusumoto, R., Yamada, A., Dohmae, N., Yokoi, M., Yuasa, M., Araki, M., Iwai, S., Takio, K., and Hanaoka, F. (1999b). The *XPV* (Xeroderma pigmentosum variant) gene encodes human DNA polymerase η. *Nature* **399**, 700–704.

McCulloch, S. D., Kokoska, R. J., Masutani, C., Iwai, S., Hanaoka, F., and Kunkel, T. A. (2004). Preferential *cis-syn* thymine dimer bypass by DNA polymerase η occurs with biased fidelity. *Nature* **428**, 97–100.

McDonald, J. P., Levine, A. S., and Woodgate, R. (1997). The *Saccharomyces cerevisiae RAD30* gene, a homologue of *Escherichia coli dinB* and *umuC*, is DNA damage inducible and functions in a novel error-free postreplication repair mechanism. *Genetics* **147**, 1557–1568.

McIlwraith, M. J., Vaisman, A., Liu, Y., Fanning, E., Woodgate, R., and West, S. C. (2005). Human DNA polymerase η promotes DNA synthesis from strand invasion intermediates of homologous recombination. *Mol. Cell* **20**, 783–792.

Moldovan, G.-L., Pfander, B., and Jentsch, S. (2007). PCNA, the maestro of the replication fork. *Cell* **129**, 665–679.

Murakumo, Y., Mizutani, S., Yamaguchi, M., Ichihara, M., and Takahashi, M. (2006). Analyses of ultraviolet-induced focus formation of hREV1 protein. *Genes Cells* **11**, 193–205.

Murakumo, Y., Ogura, Y., Ishii, H., Numata, S., Ichihara, M., Croce, C. M., Fishel, R., and Takahashi, M. (2001). Interactions in the error-prone postreplication repair proteins hREV1, hREV3, and hREV7. *J. Biol. Chem.* **276**, 35644–35651.

Nair, D. T., Johnson, R. E., Prakash, L., Prakash, S., and Aggarwal, A. K. (2005a). Human DNA polymerase ι incorporates dCTP opposite template G via a G.C + Hoogsteen base pair. *Structure* **13**, 1569–1577.

Nair, D. T., Nair, D. T., Johnson, R. E., Prakash, L., Prakash, S., and Aggarwal, A. K. (2005b). Rev1 employs a novel mechanism of DNA synthesis using a protein template. *Science* **309**, 2219–2222.

Nair, D. T., Johnson, R. E., Prakash, L., Prakash, S., and Aggarwal, A. K. (2006a). Hoogsteen base pair formation promotes synthesis opposite the $1,N^6$-ethenodeoxyadenosine lesion by human DNA polymerase ι. *Nat. Struct. Mol. Biol.* **13**, 619–625.

Nair, D. T., Johnson, R. E., Prakash, L., Prakash, S., and Aggarwal, A. K. (2006b). An incoming nucleotide imposes an *anti* to *syn* conformational change on the templating purine in the human DNA polymerase-ι active site. *Structure* **14**, 749–755.

Nair, D. T., Johnson, R. E., Prakash, S., Prakash, L., and Aggarwal, A. K. (2004). Replication by human DNA polymerase-ι occurs by Hoogsteen base-pairing. *Nature* **430**, 377–380.

Nelson, J. R., Gibbs, P. E., Nowicka, A. M., Hinkle, D. C., and Lawrence, C. W. (2000). Evidence for a second function for *Saccharomyces cerevisiae* Rev1p. *Mol. Microbiol.* **37**, 549–554.

Nelson, J. R., Lawrence, C. W., and Hinkle, D. C. (1996a). Thymine-thymine dimer bypass by yeast DNA Polymerase ζ. *Science* **272**, 1646–1649.

Nelson, J. R., Lawrence, C. W., and Hinkle, D. C. (1996b). Deoxycytidyl transferase activity of yeast *REV1* protein. *Nature* **382**, 729–731.

Ogi, T., Kannouche, P., and Lehmann, A. R. (2005). Localization of human Y-family DNA polymerase κ: Relationship to PCNA. *J. Cell Sci.* **118**, 129–136.

Ogi, T., Shinkai, Y., Tanaka, K., and Ohmori, H. (2002). Polκ protects mammalian cells against the lethal and mutagenic effects of benzo[a]pyrene. *Proc. Natl. Acad. Sci. USA* **99**, 15548–15553.

Ohashi, E., Hanafusa, T., Kamei, K., Song, I., Tomida, J., Hashimoto, H., Vaziri, C., and Ohmori, H. (2009). Identification of a novel REV1 -interacting motif necessary for DNA polymerase κ function. *Genes Cells* **14**, 101–111.

Ohashi, E., Murakumo, Y., Kanjo, N., Akagi, J., Masutani, C., Hanaoka, F., and Ohmori, H. (2004). Interaction of hREV1 with three human Y-family DNA polymerases. *Genes Cells* **9**, 523–531.

Ohashi, E., Ogi, T., Kusumoto, R., Iwai, S., Masutani, C., Hanaoka, F., and Ohmori, H. (2000). Error-prone bypass of certain DNA lesions by the human DNA polymerase κ. *Genes Dev.* **14**, 1589–1594.

Ohmori, H., Friedberg, E. C., Fuchs, R. P.P., Goodman, M. F., Hanaoka, F., Hinkle, D., Kunkel, T. A., Lawrence, C. W., Livneh, Z., Nohmi, T., Prakash, L., Prakash, S., *et al.* (2001). The Y -family of DNA polymerases. *Mol. Cell* **8**, 7–8.

Ohmori, H., Hatada, E., Qiao, Y., Tsuji, M., and Fukuda, R. (1995). *dinP*, a new gene in *Escherichia coli*, whose product shows similarity to UmuC and its homologues. *Mutat. Res.* **347**, 1–7.

Ohmori, H., Ohashi, E., and Ogi, T. (2004). Mammalian Pol κ: Regulation of its expression and lesion substrates. *Adv. Protein Chem.* **69**, 265–278.

Okada, T., Sonoda, E., Yoshimura, M., Kawano, Y., Saya, H., Kohzaki, M., and Takeda, S. (2005). Multiple roles of vertebrate *REV* genes in DNA repair and recombination. *Mol. Cell. Biol.* **25**, 6103–6111.

Pence, G. M., Blans, P., Zink, C. N., Holls, T., Fishbein, J. C., and Perrino, F. W. (2009). Lesion bypass of N^2-ethylguanine by human DNA polymerase ι. *J. Biol. Chem.* **284**, 1732–1740.

Potapova, O., Grindley, N. F., and Joyce, C. M. (2002). The mutational specificity of he Dbh lesion bypass polymerase and its implications. *J. Biol. Chem.* **277**, 28157–28166.

Prakash, S., Johnson, R. E., and Prakash, S. (2005). Eukaryotic translesion synthesis DNA polymerases: Specificity of structure and function. *Annu. Rev. Biochem.* **74**, 317–353.

Radman, M. (1999). Mutation: Enzymes of evolutionary change. *Nature* **401**, 866–869.

Rechkoblit, O., Mainina, L., Cheng, Y., Kuryavyi, V., Broyde, S., Geacintov, N. E., and Patel, D. J. (2006). Stepwise translocation of Dpo4 polymerase during error-free bypass of an 8-oxoG lesion. *PLoS Biol.* **4**, e11.

Rechkoblit, O., Zhang, Y., Guo, D., Wang, Z., Amin, S., Krzeminsky, J., Louneva, N., and Geacintov, N. E. (2002). Translesion synthesis past bulky benzo[a]pyrene diol epoxide N^2-dG and N^6-dA lesions catalyzed by DNA bypass polymerases. *J. Biol. Chem.* **277**, 30488–30494.

Reuven, N. B., Arad, G., Maor-Shoshan, A., and Livneh, Z. (1999). The mutagenesis protein UmuC is a DNA polymerase activated by the UmuD′, RecA, and SSB and is specialized for translesion replication. *J. Biol. Chem.* **274**, 31763–31766.

Ross, A. L., Simpson, L. J., and Sale, J. E. (2005). Vertebrate DNA damage tolerance requires the C-terminus but not BRCT or transferase domains of REV1. *Nucleic Acids Res.* **33**, 1280–1289.

Roush, A. A., Squarez, M., Friedberg, E. C., Radman M., and Siede, W. (1998). Deletion of the *Saccharomyces cerevisiae* gene *RAD30* encoding an *Escherichia coli* DinB homolog confers UV radiation sensitivity and altered mutability. *Mol. Gen. Genet.* **257**, 686–692.

Sabbioneda, S., Green, C. M., Bienko, M., Kannouche, P., Dikic, I., and Lehmann, A. R. (2009). Ubiquitin-binding motif in human DNA polymerase η is required for correct localization. *Proc. Natl. Acad. Sci. USA* **106**, E20.

Sabbioneda, S., Minesinger, B. K., Giannattasio, M., Plevani, P., Muzi-Falconi, M., and Jinks-Robertson, S. (2005). The 9 -1-1 checkpoint clamp physically interacts with polζ and is partially required for spontaneous polζ-dependent mutagenesis in *Saccharomyces cerevisiae. J Biol. Chem.* **280**, 38657–38665.

Shibutani, S., Takeshita, M., and Grollman, A. P. (1991). Insertion of specific bases during DNA synthesis past the oxidation-damaged base 8-oxodG. *Nature* **349**, 431–434.

Shiomi, Y., Masutani, C., Hanaoka, F., Kimura, H., and Tsurimoto, T. (2007). A second proliferating cell nuclear antigen loader complex, Ctf18-Replication Factor C, stimulates DNA polymerase η activity. *J Biol. Chem.* **282**, 20906–20914.

Silvian, L. F., Toth, E. A., Pham, P., Goodman, M. F., and Ellenberger, T. (2001). Crystal structure of a DinB family error-prone DNA polymerase from *Sulfolobus solfataricus. Nat. Struct. Biol.* **8**, 984–989.

Sohn, S. Y., and Cho, Y. (2009). Crystal structure of the human Rad9-Hus1-Rad1 clamp. *J. Mol. Biol.* **390**, 490–502.

Stelter, P., and Ulrich, H. D. (2003). Control of spontaneous and damage-induced mutagenesis by SUMO and ubiquitin conjugation. *Nature* **425**, 188–191.

Suzuki, N., Ohashi, E., Kolbanovskiy, A., Geacintov, N. E., Grollman, A. P., Ohmori, H., and Shibutani, S. (2002). Translesion synthesis by human DNA polymerase κ on a DNA template containing a single stereoisomer of dG -(+)- or dG-(–)-*anti-N*^{2}-BPDE (7,8-dihydroxy-*anti*-9,10-epoxy-7,8,9,10-tetrahydrobenzo[a]pyrene). *Biochemistry* **41**, 6100–6106.

Swan, M. K., Johnson, R. E., Prakash, L., Prakash, S., and Aggarwal, A. K. (2009). Structure of the human Rev1-DNA-dNTP ternary complex. *J. Mol. Biol.* **390**, 699–709.

Tang, M., Shen, X., Frank, E. G., O'Donnell, M., Woodgate, R., and Goodman, M. F. (1999). UmuD′$_2$C is an error-prone DNA polymerase, *Escherichia coli* pol V. *Proc. Natl. Acad. Sci. USA* **96**, 8919–8924.

Tissier, A., Kannouche, P., Reck, M. P., Lehmann, A. R., Fuchs, R. P., and Cordonnier, A. (2004). Co-localization in replication foci and interaction of human Y-family members, DNA polymerase pol η and REVl protein. *DNA Repair (Amst)* **3**, 1503–1514.

Tissier, A., McDonald, J. P., Frank, E. G., and Woodgate, R. (2000). polι, a remarkably error-prone human DNA polymerase. *Genes Dev.* **14**, 1642–1650.

Trincao, J., Johnson, R. E., Escalante, C. R., Prakash, S., Prakash, L., and Aggarwal, A. K. (2001). Structure of the catalytic core of *S. cerevisiae* DNA polymerase η: Implications for translesion DNA synthesis. *Mol. Cell* **8**, 417–426.

Tsuji, Y., Watanabe, K., Araki, K., Shinohara, M., Yamagata, Y., Tsurimoto, T., Hanaoka, F., Yamamura, K., Yamaizumi, M., and Tateishi, S. (2008). Recognition of forked and single-stranded DNA structures by human RAD18 complexed with RAD6B protein triggers its recruitment to stalled replication forks. *Genes Cells* **13**, 343–354.

Uljoin, S. N., Johnson, R. E., Edwards, T. A., Prakash, S., Prakash, L., and Aggarwal, A. K. (2004). Crystal structure of the catalytic core of human DNA polymerase κ. *Structure* **12**, 1395–1404.

Vaisman, A., Lehmann, A. R., and Woodgate, R. (2004). DNA polymerases η and ι. *Adv. Protein Chem.* **69**, 205–228.

Vidal, A. E., Kannouche, P., Podust, V. N., Yang, W., Lehmann, A. R., and Woodgate, R. (2004). Proliferating cell nuclear antigen-dependent coordination of the biological functions of human DNA polymerase ι. *J Biol. Chem.* **279**, 48360–48368.

Vidal, A. E., and Woodgate, R. (2009). Insights into the cellular role of enigmatic DNA polymerase ι. *DNA Repair* **8**, 420–423.

Wagner, J., Gruz, P., Kim, S. R., Yamada, M., Matsui, K., Fuchs, R. P., and Nohmi, T. (1999). The *dinB* gene encodes a novel *E. coli* DNA polymerase, DNA PolIV, involved in mutagenesis. *Mol. Cell* **4**, 281–286.

Warbrick, E. (1998). PCNA binding through a conserved motif. *Bioessays* **20**, 195–199.

Ward, J. J., Sodhi, J. S., McGuffin, L. J., Buxton, B. F., and Jones, D. T. (2004). Prediction and functional analysis of native disorder in proteins from the three kingdoms of life. *J. Mol. Biol.* **337**, 635–645.

Watanabe, K., Tateishi, S., Kawasuji, M., Tsurimoto, T., Inoue, H., and Yamaizumi, M. (2004). Rad18 guides polη to replication stalling sites through physical interaction and PCNA monoubiquitination. *EMBO J.* **23**, 3886–3896.

Weill, J.-C., and Reynaud, C. -A. (2008). DNA polymerase in adaptive immunity. *Nat. Rev. Immunol.* **8**, 302–312.

Wilson, R. C., and Pata, J. D. (2008). Structural insights into the generation of single-base deletions by the Y-family DNA polymerase Dbh. *Mol. Cell* **29**, 767–779.

Wilson, T. M., Vaisman, A., Martomo, S. A., Sullivan, P., Lan, L., Hanaoka, F., Yasui, A., Woodgate, R., and Gearhart, P. J. (2005). MSH2-MSH6 stimulates DNA polymerase η, suggesting a role for A:T mutations in antibody genes. *J. Exp. Med.* **201**, 637–645.

Wong, J. H., Fiala, K. A., Suo, Z., and Ling, H. (2008). Snapshots of a Y-family DNA polymerase in replication: Substrate-induced conformational transitions and implications for fidelity of Dpo4. *J. Mol. Biol.* **379**, 317–330.

Xing, G., Kirouac, K., Shin, Y. J., Bell, S. D., and Ling, H. (2009). Structural insight into recruitment of translesion DNA polymerase Dpo4 to sliding clamp PCNA. *Mol. Microbiol.* **71**, 678–691.

Xu, M., Bai, L., Gong, Y., Xie, W., Hang, H., and Jiang, T. (2009). Structure and functional implications of the human Rad9-Hus1-Rad1 cell cycle checkpoint complex. *J. Biol. Chem.* **284**, 20457–20461.

Yang, W., and Woodgate, R. (2007). What a difference a decade makes: Insights into translesion synthesis. *Proc. Natl. Acad. Sci. USA* **104**, 174–183.

Yuasa, M. S., Masutani, C., Hirano, A., Cohn, M. A., Yamaizumi, M., Nakatani, Y., and Hanaoka, F. (2006). A human DNA polymerase η complex containing Rad18, Rad6 and Rev1; proteomic analysis and targeting of the complex to the chromatin-bound fraction of cells undergoing replication fork arrest. *Genes Cells* **11**, 731–744.

Zang, H., Irimia, A., Choi, J. Y., Angel, K. C., Loukachevitch, L. V., Egli, M., and Guengerich, F. P. (2006). Efficient and high fidelity incorporation of dCTP opposite 7,8-dihydro-8-oxodeoxyguanosine by *Sulfolobus solfataricus* DNA polymerase Dpo4. *J. Biol. Chem.* **281**, 2358–2372.

Zhang, H., Gibbs, P. E.M., and Lawrence, C. W. (2006). The *Saccharomyces cerevisiae rev6-1* mutation, which inhibits both the lesion bypass and the recombination mode of DNA damage tolerance, is an allele of POL30, encoding proliferating cell nuclear antigen. *Genetics* **173**, 1983–1989.

Zhang, Y., Yuan, F., Wu, X., Wang, M., Rechkoblit, O., Taylor, J. S., Geacintov, N. E., and Wang, Z. (2000). Error-free and error-prone lesion bypass by human DNA polymerase κ. *Nucleic Acids Res.* **28**, 4138–4146.

Zhou, B.-L., Pata, J. D., and Steitz, T. A. (2001). Crystal structure of a DinB lesion bypass DNA polymerase catalytic fragment reveals a classic polymerase catalytic domain. *Mol. Cell* **8**, 427–437.

AUTHOR INDEX

SUBJECT INDEX

NOTE: The letters 'f' and 't' following the locators refer to figures and tables respectively.

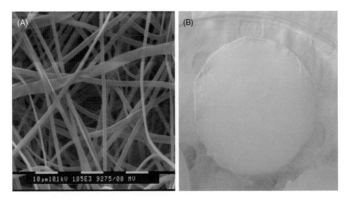

Wise *et al.*, Chapter 1, Fig 2.   Recombinant human elastin can be electrospun to form (A) fine microfibers that can in turn accumulate to form (B) scaffolds for biomaterials applications.

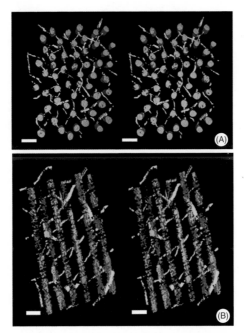

Knupp *et al.*, Chapter 2, Fig. 7.   Three-dimensional reconstruction from the central anterior region of mouse cornea. (A) Transverse view. (B) Longitudinal view. Collagen fibrils are painted in blue and proteoglycans in yellow. The proteoglycans shown in the reconstructions are mainly those with chondroitin/dermatan sulfate glycosaminoglycan chains. No apparent regular proteoglycan labeling of the collagen fibrils is visible. Bars = 50 nm.

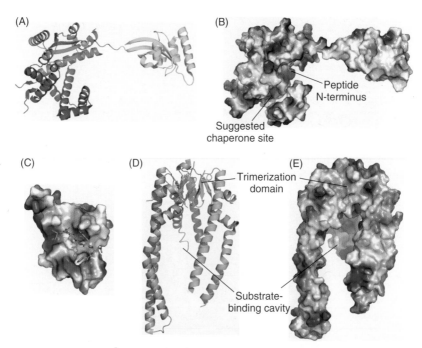

ALLEN *ET AL.*, CHAPTER 3, FIG 1.   Structures of outer membrane protein chaperones SurA and Skp (PDB code 1SG2). (A) Cartoon representation of full-length SurA [PDB code 1M5Y (Bitto and McKay, 2002)], with the N-terminal domain colored blue, the P1 domain green, the P2 domain yellow, and the C-terminal domain red. (B) Surface model of full-length SurA in the same orientation as in (A), colored by a gradient of electrostatic potential from negative (red) to positive (blue). A substrate peptide (WEYIPNV) bound to the P1 pocket is shown in stick representation with carbons in green, nitrogens in blue, and oxygens in red. The orientation of this peptide was determined by superposition of the crystal structure of the P1 domain with bound peptide [PDB code 2PV1 (Xu *et al.*, 2007)] onto the P1 domain of the full-length structure. (C) Surface model of the P1 domain of SurA in complex with the model peptide, using the same display mode as in (B). (D) Cartoon structure of Skp [PDB code 1SG2 (Walton and Sousa, 2004)], with the three subunits forming the trimer colored in cyan, magenta, and orange, respectively. (E) Surface model of Skp in the same orientation as in (D), colored by a gradient of electrostatic potential from negative (red) to positive (blue).

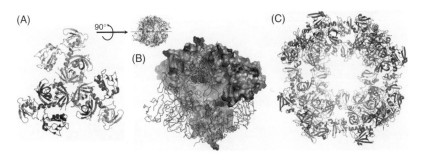

ALLEN *ET AL.*, CHAPTER 3, FIG 2. Structures of DegP in various assembly states. (A) Cartoon representation of a DegP trimer from the original hexamer structure [PDB code 1KY9 (Krojer *et al.*, 2002)], viewed from above with the protease domains colored red, the PDZ1 domains in blue, and the PDZ2 domains in green. (B) Side view of the DegP hexamer, with three of the subunits (two from the trimer at the top and one from the trimer at the bottom) shown colored by a gradient of electrostatic potential from negative (red) to positive (blue). The remaining three subunits are shown in backbone ribbon representation, using the same color scheme as in (A). (C) Twelve of the DegP subunits from the 24-mer crystal structure [PDB code 3CS0 (Krojer *et al.*, 2008)], looking into the inside of the cavity and shown using the same display mode as (A).

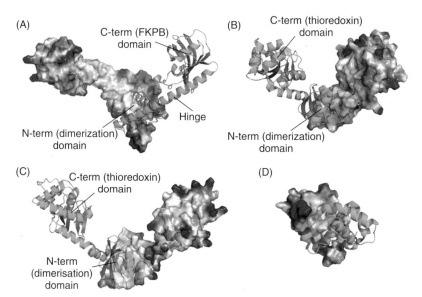

ALLEN *ET AL.*, CHAPTER 3, FIG 3.   Structures of various soluble chaperones involved in general purpose protein folding in the periplasm. In each case, the physiological dimers are shown with one subunit in surface representation, colored by a gradient of electrostatic potential from negative (red) to positive (blue), and the other subunit in cartoon representation, colored by secondary structure with α-helices in cyan, β-sheets in magenta, and turn regions in pink. (A) FkpA [PDB code 1Q6U (Saul *et al.*, 2004)]. In the chosen orientation, one of the three α-helices in the dimerization domain of the cartoon structure (helix 2) is obscured by the end of helix 1 in the surface structure. This is because the two domains loop through each other, giving rise to the robust dimer interface. (B) DsbC [PDB code 1EEJ (McCarthy *et al.*, 2000)]. (C) DsbG [PDB code 1V57 (Heras *et al.*, 2004)]. (D) HdeA [PDB code 1DJ8 (Gajiwala and Burley, 2000)]

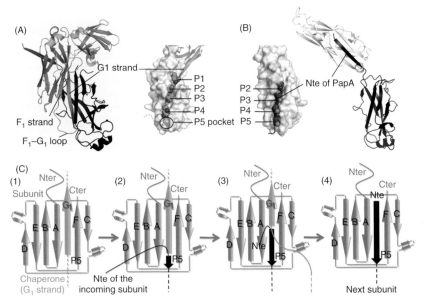

ALLEN ET AL., CHAPTER 3, FIG 4.   The chaperone/usher pathway. (A) Two different views of the archetypal chaperone PapD in complex with the pilus rod subunit PapA [PDB code 2UY6 (Verger et al., 2007)]. Left panel: cartoon representation of a PapA subunit (colored black) in donor-strand complementation (DSC) with the PapD chaperone (colored by secondary structure with α-helices in cyan, β-sheets in magenta, and turn regions in pink). The β-strands F_1 and G_1 and the F_1–G_1 loop between them are indicated. Right panel: surface representation of the PapA subunit with the G_1 strand of the chaperone shown in cartoon representation. The chaperone's residues P1–P4 bound within the P1–P4 sites/regions of the subunit's hydrophobic binding groove are highlighted in space-filling representation (carbon in green, oxygen in red, and nitrogen in blue). (B) Two different views of two pilus subunits in donor-strand exchange (DSE) [PDB code 2UY6 (Verger et al., 2007)]. For clarity, the chaperone associated with the first subunit is not shown. Right panel: cartoon representation with one subunit in white and the other in black. Left panel: surface representation of the subunit in white at right, with the Nte of the subunit in black at right shown as a cartoon. Nte's residues P2–P5 bound within the P2–P5 sites/ regions of the hydrophobic binding groove are highlighted in space-filling representation (carbon in black, oxygen in red, and nitrogen in blue). (C) Schematic diagram of the DSC and DSE mechanisms. (1) Topology diagram of the pilus subunit (in gray) showing the Ig-like fold complemented by the insertion of the G_1 strand from the chaperone (in green). In this donor-strand-complemented conformation, the pocket P5 is unoccupied. (2) Initial positioning of the incoming subunit's N-terminal extension (Nte, in black) via insertion of the P5 residue of the incoming subunit's Nte into the P5 pocket of the receiving subunit's groove. (3) Progressive insertion of the incoming subunit's Nte inside the receiving subunit's groove. While the Nte zips in, the chaperone's G1 strand zips out. (4) Completed DSE reaction. The chaperone is now completely released, and the receiving pilus subunit is stabilized by the completed insertion of the Nte from the incoming subunit.

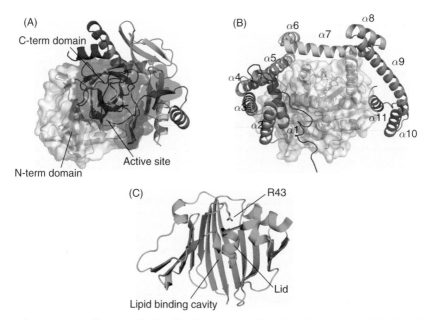

ALLEN *ET AL.*, CHAPTER 3, FIG 5. Structures of carrier chaperones. (A) Crystal structure of α-lytic protease (αLP) with its prodomain chaperone [PDB code 4PRO (Sauter *et al.*, 1998)]. The prodomain is in cartoon representation colored in rainbow and the αLP is in both cartoon/surface representations. The N- and C-terminal domains of αLP are colored, respectively, in light gray and black. The location of the active site of LP is indicated. It is occupied by the insertion of the prodomain C-terminal tail in red. (B) Crystal structure of lipase A (LipA) bound to its chaperone Lif [PDB code 2ES4 (Pauwels *et al.*, 2006)]. The chaperone is in cartoon representation and colored in rainbow. LipA is in cartoon and surface representations, colored in gray. (C) Structure of LolA [PDB code 1IWL (Takeda *et al.*, 2003)] shown in cartoon representation with α-helical regions in cyan, β-sheet in magenta, and loop regions in pink. Arginine 43 is marked out in stick representation with nitrogen atoms colored in blue.